ECO-NOMICS

RICHARD L. STROUP

ECO-NOMICS

WHAT EVERYONE SHOULD KNOW ABOUT ECONOMICS AND THE ENVIRONMENT

CATO
INSTITUTE
WASHINGTON, D.C.

Library of Congress Cataloging-in-Publication Data

Stroup, Richard.
  Eco-nomics: what everyone should know about economics and the
environment / Richard L. Stroup.
     p. cm
  Includes bibliographical references and index.
  ISBN 1-930865-44-9 (pbk. : alk. paper)
     1. Environmental economics. 2. Environmental policy–Economic
aspects. I. Tilte.

  HC79.E5S775 2003
  333.7–dc21

                                                         2003043783

Paperback ISBN: 978-1-944424-00-8
eBook ISBN: 978-1-944424-01-5

Cover design by Jon Meyers.
Printed in Canada.

                          CATO INSTITUTE
             1000 Massachusetts Ave., N.W.
                  Washington, D.C. 20001
                       www.cato.org

Contents

List of Figures

Acknowledgments

I am pleased that a new edition of *Eco-nomics* is being published with additional and updated information. Although the book is a small one, I have attempted to identify in it the core tenets of free-market approaches to environmental protection and to make clear why these approaches are worth serious consideration.

I am grateful to many people who have helped me develop and write about these ideas and examples, beginning with John Baden of the Foundation for Research on Economics and the Environment and with Terry Anderson and P. J. Hill of PERC, the Property and Environment Research Center. My years working with PERC helped me hone the concepts and examples, and many of the staff and visiting scholars at PERC made solid contributions to the development of what we now call free-market environmentalism. I appreciate the contributions of all.

And, of course, I owe a great debt to James D. Gwartney, with whom I have worked for many years, most notably on our principles text, *Economics: Private and Public Choice*, now in its 15th edition (and now with coauthors Russell S. Sobel and David A. Macpherson).

I would like to thank Matthew Brown, now executive director of the Academy on Capitalism and Limited Government at the University of Illinois, and Sandra Goodman, formerly a PERC fellow. Both worked with me on the contents of the previous edition, as did my wife, Jane Shaw Stroup, who had a major role in preparing this edition.

Thanks are due to the Cato Institute, specifically to Jerry Taylor, who recognized the value of the original edition, and to John Samples, who proposed that it be updated and expanded and who helped me accomplish it.

1. Changing Environmentalism

The year was 1984. As he walked through the door of a small schoolhouse in Saint Anthony, Idaho, Hank Fischer, a young environmentalist who worked for a group called Defenders of Wildlife, faced a "sea of cowboy hats." He had come to the meeting to persuade local ranchers to allow wolves to be brought back to Yellowstone National Park and central Idaho. Wolves had been exterminated in Yellowstone and other parts of the Mountain West decades earlier.

As Fischer started his talk, a sheep rancher he had met earlier stood up and called out, "Hank Fischer, you mean nobody's kilt you yet?" Fischer recognized that this was an attempt at humor, but it was unnerving.

Fischer quickly learned that ranchers were not going to accept wolves in their region unless he could change their incentives. So he began thinking outside the box. He persuaded donors to Defenders of Wildlife to create a wolf compensation program.

And that made the difference. Once he started compensating ranchers for livestock that had been killed by wolves, "the wolf/livestock conflict was no longer an issue dominating the newspapers," he wrote later. "It disappeared. I went back to my organization and said, 'Let's keep doing this.'"[1]

Fischer went on to use compensation to protect grizzly bears as well as wolves. He is now employed by the National Wildlife Federation and works with ranchers to retire grazing land in the Yellowstone region so that more wildlife can thrive.

Fischer is an unusual environmentalist because he recognizes that people are part of nature and that their views and incentives matter. He didn't know it at the time, but he was looking at wildlife introduction as an economist would.

This book is about applying economics to help us understand and solve environmental issues. But it's important to understand just what environmentalism is, because it is a constantly changing landscape. That is the purpose of this chapter.

Fifty years before Hank Fischer walked through the Idaho schoolhouse door, Rosalie Barrow Edge was making people uncomfortable—and this time they were environmentalists, not ranchers. An affluent divorced matron who lived in New York in the early and mid-20th century, Rosalie Edge was an activist for women's suffrage in her 30s and an activist for the environment in her 50s.[2]

Living across town from New York's magnificent Central Park, she rediscovered a childhood love of bird watching. She began walking through the park, carrying a notebook, and listening and looking for birds. She was entranced. She enlisted the curiosity of her 12-year-old son, and they began watching birds together.

Inevitably, Edge joined the National Association of Audubon Societies, the leading society for protecting birds. But she soon became a thorn in its side.

The association of Audubon societies had been formed in 1887 to bring together bird enthusiasts who had organized local Audubon societies. Its chief goal was to reverse the loss of beautiful creatures like herons and egrets, whose plumes were being used in women's hats. Indeed, one of the societies' activities was to campaign against the purchase of feathers for women's hats and clothing.

But the Audubon Society didn't want to protect *all* birds.

Edge's problem with the Audubon Society started when she came across an inflammatory pamphlet, "Crisis in Conservation." It castigated unnamed "bird protection organizations" for failing to oppose bounties (such as Alaska's bounty on the bald eagle, the national bird) and actually undermining the goal of bird protection. (The organizations were unnamed, but there was no doubt that this pamphlet was directed at Audubon.) Edge learned that the association had entered into an agreement with gun manufacturers: the society could save all the birds it wanted as long as it didn't oppose sport hunting.

So Edge started a campaign against bird hunting. She became unpopular with the wealthy leaders of the Audubon association. She was even chastised by Edith Roosevelt, the widow of Theodore Roosevelt. In a letter signed by other dignitaries as well, Mrs. Roosevelt deplored the "small group of people" in the association who were undermining its leadership. But Edge had secret allies who cheered her on privately even though they were unwilling to take a public stand.

Although Edge didn't win all her fights with Audubon, she did succeed in protecting birds of prey. Her most acclaimed achievement was the creation of the Hawk Mountain Sanctuary. Disturbed that hunters gathered at Hawk Mountain in eastern Pennsylvania to slaughter hawks by the thousands, she decided to lease the mountain and keep hunters out. The private preserve, founded in 1934, still exists today and is known for its raptor research.

Like Hank Fischer, Rosalie Edge was an environmental innovator who epitomized the constantly changing view of our environment and how to protect it.

Indeed, differences—and sometimes conflicts—over how to protect the environment go back at least to the late 19th century. To John Muir, an early American naturalist and geologist, the dramatic landscape of the California Sierras was sacred. As Robert H. Nelson has written about Muir, "Of the Sierra wilderness, Muir declared that 'everything in it seems equally divine—one smooth, pure, wild glow of Heaven's love.'"[3] Muir founded the Sierra Club and spent much of his later life promoting the idea of national parks so that those landscapes could remain pristine.

At about the same time, Gifford Pinchot, an affluent young Yale graduate and forester, was developing a different concept of environmental protection. Late in the 19th century, many people had become alarmed about a "timber famine." So many trees had been cut down across America, especially wide swaths of forest in the northern Midwest states of Wisconsin and Michigan, that some thought forests would disappear. Pinchot was worried, too. He wanted the government to protect the trees—while still selling them commercially. It did. Pinchot became the first director of the U.S. Forest Service.

The split between environmental organizations that promoted useful "conservation" (as Pinchot did) and those that wanted pristine "preservation" (as Muir did) continues today. It illustrates that there is no single "environmentalist" position on many matters.

Environmentalists were active during much of the 20th century, but the modern environmental movement—something of a mass movement—took hold in the 1960s.

Some claim that it started with the opposition that developed over the Storm King hydropower project in New York State. That project, proposed in 1963 by a private power company, Con Edison, would have pumped water from the Hudson River to generate energy and distribute it throughout the region. The opponents argued that the project, with its pipes, reservoir, and transmission lines, would violate

3

the natural beauty of the Hudson Valley and Storm King Mountain. Ultimately, they prevailed. The case set important legal precedents and led to the first modern environmental law, the National Environmental Policy Act, signed by President Richard Nixon on January 1, 1970.

Others, however, give more credit to Rachel Carson, whose 1962 book *Silent Spring* recorded the extensive use of the chemical DDT to wipe out mosquitoes and other insects. She argued that DDT had led to the obliteration of birds, especially peregrine falcons, by thinning their eggshells. Although the issue was debated for years—and to some extent still is—the Environmental Protection Agency (EPA) banned most uses of DDT in 1972. The decision by EPA administrator William Ruckelshaus showed the power of the federal government in the environmental arena and launched the Environmental Protection Agency as a major government force.

Meanwhile, young people who had cut their teeth protesting against nuclear weapons and the Vietnam War had turned their attention to air and water pollution.

Such pollution had occurred frequently in the past, but court suits and local regulation had helped to constrain it. (So had new technology: in spite of its emission of pollutants, the automobile had improved the urban environment because horse manure no longer muddied the streets or spread disease.) But in the 1960s, the new wealth of the country, with its affluent college-educated young people eager to make everything right, led to massive efforts, local and national, to clean up air and water.

Emblematic of the effort was the original Earth Day, held on April 22, 1970, spurred by Wisconsin senator Gaylord Nelson, who became a champion of environmental legislation. According to *National Geographic*, 20 million Americans took part in that day's speeches and demonstrations around the country.[4]

Soon health issues, especially fear of cancer caused by industrial chemicals, surfaced more forcefully than in the past. And then there were the ugliness of mining, worries about pollution of the ocean, and alarm over nuclear power. From 1969 to 1980, about a dozen laws were enacted that gave the federal government authority to control pollution. They took their place alongside earlier laws, such as the Migratory Bird Treaty Act and the National Park Service Organic Act, which reflected more traditional environmental concerns.

The next 50 years were punctuated by dramatic environmental events, some of them horrendous, such as the 1984 leak of toxic gas in Bhopal, India, which killed thousands of people, and the 1986 Chernobyl nuclear power plant accident in the Soviet Union, which killed 31 people and contaminated many more with radiation. But few events like that have occurred in the United States.

The closest the United States has come to an environmental disaster on that scale has been through notorious oil spills such as the Exxon spill of 1989 off the coast of Alaska and the lethal BP oil spill of 2010 in the Gulf of Mexico. The latter was caused by an explosion that killed 11 workers and caused oil to gush for 87 days. Those spills, however, did not destroy species and seem to have had little long-term environmental impact.

Most of the apparent environmental crises in the United States over the past half-century have turned out to be exaggerated.

- The Love Canal incident, which made headlines for months and led to a tremendous expansion of the EPA, is largely forgotten now. The 1976 discovery of a buried waste dump in Niagara Falls, New York, oozing chemicals and located next to a school, created fear that the nation was covered with hazardous waste dumps that were "ticking time bombs." But the fears were much exaggerated.

- An accident at a nuclear power plant in Pennsylvania, Three Mile Island—caused partly by human error—magnetized public attention for weeks as it turned into a dreaded partial meltdown. Yet no one was killed or injured.

- For years, the public and scientists alike feared that lakes and forests in the eastern United States were being destroyed by acid rain caused by sulfur from electric power plants that killed fish and trees. Yet a 10-year, $500 million government study issued in 1990 concluded that at most a few high-altitude red spruce forests, damaged by cold weather and wind, may have been harmed by acid rain. Many of the lakes, it turned out, are naturally acidic.

Today, environmental issues are far different from those even 30 years ago. In part, that's because visible pollution has almost disappeared—the plumes from power plants are mostly steam (and carbon dioxide, today's bête noire, is invisible). Fears of toxic pollution have largely subsided.

Nevertheless, the impetus to protect the environment remains.

It is important to keep North American environmental problems in perspective, however. In 1990, James R. Dunn and John E. Kenney compared two lists of environmental problems: one covering the United States, one covering Africa. Americans had many environmental worries, from hazardous waste sites to destruction of the ozone layer. For Africans, the list consisted of diseases such as sleeping sickness and cholera, soil erosion, and lack of sewage treatment.

Dunn and Kenney called the American list a "media list in the sense that the public must be told about most problems (that is, most citizens do not really see or feel the problems on a daily basis)." In contrast, the African perils were "Third World megaproblems—noncontroversial, pervasive, and highly visible."[5]

Today's big environmental issue, at least in the United States, is climate change. This is the current term for the concern about global warming that emerged in the late 1980s. Some scientists contend that the world is heading for a cataclysmic impact from rising temperatures. Their proposed solution, picked up by many politicians, is to reduce the use of fossil fuels because the burning of those fuels emits carbon dioxide, a gas that traps heat that would otherwise be radiated into space.

Second only to climate change today is the somewhat enigmatic and stretchable concept "sustainability." It stems from *Our Common Future*, a 1987 report by a United Nations commission headed by Grø Brundtland, then prime minister of Norway. The commission wanted to promote "sustainable development," which is defined in the report as development that "meets the needs of the present without compromising the ability of future generations to meet their own needs."[6]

This upbeat statement has been interpreted by some to mean that there should be no tampering with natural habitat or conditions, and it has led to a motley collection of proposals ranging from creating coastal fish preserves to recycling everything that can possibly be recycled. In fact, on college campuses, "sustainability" sometimes sweeps in social justice and related concepts.

So the history of environmentalism includes a constant shift from one concern to another. What ties the concerns together? For the most part, it is a desire to keep what surrounds us beautiful, clean, valuable—and "natural." One way to view environmentalism is to see it as what lies outside our own property—outside the things we have personal control over and, in many cases, outside the bounds of what *anyone* has personal control over.

In this book, as indicated, we will apply economics to environmentalism. Economics is the study of people's choices under different incentives. It will help us understand why some of our surroundings are pristine, why some are polluted, and why many are somewhere in between. And it can help guide us in making the best environmental choices going forward.

Economic principles explain why we cannot have everything we want but instead must make tradeoffs, how people respond to changed incentives, and how the institutions of our country and society—the rules, customs, and habits that guide us—will affect our ability to protect the environment.

The next four chapters will introduce these principles and illustrate them with various ways that people have dealt with (or failed to deal with) environmental issues. The final chapter (Chapter 6) will apply those principles to the most talked-about issue of the day, climate change, and its connection with energy.

2. Scarcity: An Economics Primer

This chapter introduces 10 principles of economics that shed light on environmental problems. They are presented as the answers to commonly asked questions about environmental issues. It should quickly become clear that economics is about choice, not necessarily about money, and that economics can help us understand environmental choice, both public and private.

1. In a land as rich as the United States, why do we face so many difficult choices about the environment?

Scarcity, even in a nation as wealthy as the United States, is always with us, so choices must be made.

We have vast forests in this country but not enough to provide all of the wood, all of the wilderness, and all of the accessible recreation that we want. As soon as we log trees, build roads, or improve trails and campsites, we lose some wilderness. Similarly, we have large amounts of fresh water, but if we use water to grow rice in California, the water consumed cannot be used for drinking water in California cities. If we use fire to help a forest renew itself, we will have air pollution downwind while the fire burns. We have many goals, so we have to make choices about how to allocate our limited resources. The cost of those choices is what we give up—the cost of opportunities lost.

The trouble is that people have differing goals and disagree about which choice is the best one. Pursuit of differing goals may lead to conflict. Nowhere is that situation clearer than in environmental matters.

In the mid-1990s, California's San Bernardino County was about to build a new hospital. Less than 24 hours before groundbreaking, the U.S. Fish and Wildlife Service announced that the flower-loving Delhi Sands fly, which had been found on the site, was an endangered species. So the county had to spend $4.5 million to move the hospital 250 feet to give the flies a few acres to live on and a corridor to the nearby sand dunes. It also had to divert funds from its medical mission to pay for biological studies of the fly.[1]

9

Environmentalists, who wanted biological diversity, were relieved that the hospital would move, but county officials were upset at the delay and the high cost that their hospital budget and the taxpayers would have to bear. To use resources one way sacrifices the use of those resources for other things. There is no escaping cost.

San Bernardino County faced a choice between timely provision of a health care facility and protection of a unique species. Often the choices, however, are between different environmental goals. For example, many environmentalists consider wind energy a desirable source of energy because it doesn't use coal. However, the giant structures that harvest the wind energy do kill birds. In fact, the U.S. Fish and Wildlife Service reported in 2013 that 85 gold and bald eagles—both protected species—were killed by flying into windmills between 1997 and 2012.[2]

And in 2011, several organizations, including a conservation group, brought a suit against the builders of five solar thermal plants in desert areas in southern California. The plaintiffs argued that because the facilities use a lot of water, they are endangering the surrounding land. Those are "fragile landscapes," the *New York Times* reported, and "home to desert tortoises, bighorn sheep and other protected flora and fauna."[3]

Either of these choices may be defensible, but it's necessary to choose. Scarcity is a fundamental fact of life, not just of economics. It is always present in nature, even when human beings are not. Each population of a species can flourish and expand only until it reaches the limit of available habitat, sunlight, water, and nutrients—all of which are scarce. Trees grow taller as they compete for the limited amount of sunlight that nourishes them. Some plants spread their leaves horizontally, capturing sunlight—but also blocking access for other species that might sprout up to compete for water and nutrients. Each successful strategy captures resources, taking them from some competing species populations.

The notion of competition implies that some species will lose out. This loss can happen slowly over time as change occurs. When a niche in the habitat changes, for example, each population, using a different strategy, gains or loses relative to its competitors.

Even small changes in a habitat can reallocate space, water, and nutrients among populations of various species and can change the competitive outcome. Every change in a local environment will favor some species at the expense of others. And local environments are always changing over time, whether humans are present or not.

In other words, scarcity and competition are not ideas that are limited to selfish human beings.

2. Even though economists emphasize selfish motives, don't people have common goals? Doesn't everyone want a safe and attractive environment?

People share many values, but each person has a narrow focus and somewhat different purposes; each person wants to emphasize different goals.

The goals of some individuals may be selfish—intended to further only their own welfare. The goals of others are often altruistic—intended to help fellow human beings. In both cases, each person's concern and vision are focused mainly on a narrow set of ends.

Even the most noble and altruistic goals are typically narrow. Consider a couple of famous examples. The late Mother Teresa, one of the most admired women in the world, was legendary for her service to the indigent and the sick of Calcutta. Also admired was Sierra Club founder John Muir, whose vision of pristine nature was discussed in the first chapter. In both cases, their goals were widely regarded as noble and altruistic, not narrowly selfish.

Yet one is tempted to assume that Mother Teresa would have been willing to sacrifice some of the remaining wilderness in India to provide another hospital for the people she cared so much about—those dying in Calcutta. And John Muir would have been willing to see fewer hospitals if that approach helped preserve wilderness. All individuals, whether selfish or not, are narrowly focused. Each individual is willing to see sacrifices made in other less important goals to further his or her own narrow purposes.

As Adam Smith, the founder of classical economics, pointed out more than 200 years ago, we know and care most about things that directly affect us, our immediate family, and others close to us. We know much less about things that mostly affect people we never see. When people further their own set of goals, it doesn't mean that they care nothing about others. It just means that for each of us, our strongest interests are narrowly focused. These narrow goals, whatever the mix of selfishness and altruism, correspond to what economists call the "self-interest" of an individual.

It is unavoidable that an individual's choices will be driven by a limited focus. Thus, people who call themselves environmentalists may differ from others who place a higher priority on providing good schools or hospitals or making sure that poor people are well provided for.

And they may also differ on which environmental goals to pursue. There are thousands of worthy environmental goals, but each competes with others for our limited land, water, and other resources. Even without selfishness, the focused goals of individuals are enough to ensure that there will be strong disagreements and competition for scarce natural resources.

This narrowness of emphasis is important for understanding the economics of environmental issues. Depending on the circumstance, such goals can lead to tunnel vision, with destructive results, or to satisfying exchanges that make all participants better off.

3. Why do fierce arguments between organizations and individuals erupt over decisions about our resources and environment?

Although scarcity guarantees competition, some forms of competition lead to constructive action that reduces scarcity, whereas other forms are destructive.

Disagreement on values is normal. Some environmentalists who strongly appreciate the protection of fish plus the recreational and aesthetic benefits of wild, free-flowing rivers are trying to prevent the construction of dams in Chile, India, and the United States—and they are hoping to remove some that already exist. But others value the flood protection, recreation, and clean hydropower provided by the dams. Similarly, those individuals who want vast stretches of wild land lobby to prevent the construction of new roads in roadless areas, whereas people who want greater public access to the same lands lobby for additional roads and campgrounds.

The same lands and rivers cannot simultaneously provide the advantages of preservation in a wild state and the benefits of development. Competition over the management of these rivers and lands is inevitable. The only question is the form that competition will take.

Human competition can be violent, or it can be peaceful and constructive. Markets are generally peaceful because exchanges are voluntary. Even the repellent term "cutthroat competition" refers to a constructive activity: it means offering buyers low prices to get them to buy something. Sellers compete for buyers by improving their products and lowering their costs.

Human competition can also be destructive. Wars are the prime example, of course, but competition can be destructive even when it is not violent. Political battles, for example, can result in costly and unpleasant smear campaigns by various sides, each seeking to take votes from the other.

4. As people seek to meet their goals, can we predict how they will choose among the many ways they can advance those goals?

Yes.

Incentives matter.

Nearly everyone would want to save a person who is drowning. But each of us is more likely to try to rescue a person who falls into two feet of water at the edge of a small pond than to try to rescue someone who falls into the water above Niagara Falls. In other words, whatever the goal, we can predict that people will more likely act to achieve it when the cost to them is small, and they will seek low-cost ways—low cost to themselves and to their goals—to do so. These costs and benefits—or penalties and rewards—are called "incentives."

Incentives help us to understand behavior. If a person's goal is to increase income, that person has an incentive to devote long hours to work, perhaps even to a grueling job. If federal taxpayers can help pay the cost of a highway in one state, the state legislature has an additional incentive to build the highway. If people can protect an endangered species without disrupting their lives, they are more likely to choose to save it.

Incentives also affect the methods people use to achieve a particular goal. The history of the aluminum can, for example, illustrates how incentives change the use of materials. In the 1970s, as the production of aluminum became more costly, engineers figured out how to make cans thinner and thinner. According to two engineers writing in 1994, companies could save $20 million a year for each 1 percent reduction in the amount of aluminum per can.[4] That was a significant incentive to make aluminum cans lighter but just as protective.

It is not difficult for us as individuals to recognize and evaluate the cost of different choices. We are well tuned to the relative costs that we personally face when considering alternatives. However, it's more difficult to recognize and take into account the costs facing others. Costs to others will have less effect on our choices than the costs—and benefits—that we incur directly will have. This is simply the result of our narrow focus, already mentioned.

13

Typically, we expect people in business or individuals seeking personal goals to be more sensitive to their own costs than to those of others. We sometimes assume that government officials will not be so self-interested. But this assumption is not so, as economists in the field of public choice have shown. One illustration is a well-known court case brought by South Carolina developer David Lucas. It shows that officials of South Carolina were more sensitive to their own costs than to those of their constituents.[5]

The saga began when the state passed a law regulating construction along its coastline, presumably to preserve open space and to prevent possible erosion. David Lucas owned two lots along the shore, but once the law was passed, officials told him that he could not build there, even though people next to his property had already built homes on their shoreline properties. Lucas lost nearly all the value of his land.

Lucas believed that if the state wanted to control his land for a public purpose (other than stopping him from harming other people or property), the state should pay for it. So he sued to force payment. Initially, Lucas lost, but he appealed all the way to the U.S. Supreme Court and finally won. The court told South Carolina that it must pay for the land because it had taken from Lucas the same rights to use it that his neighbors enjoyed.

Once the state had to pay Lucas more than $1 million, officials changed their minds about keeping the land from development. In fact, the state sold the land to a developer!

Earlier, when they thought Lucas would pay the cost of stopping development, state regulators had little incentive to worry about the cost. But when forced to bear the cost from their own budget, they made the opposite decision: they allowed development. Incentives mattered.

The Endangered Species Act also illustrates the harm that can occur when one party (in this case, the government) determines how another (in this case, landowners) must use land. Under the act, government officials have great latitude in telling landowners what to do if they find an endangered animal such as a red-cockaded woodpecker on their property. The government chooses the protection methods, but the landowner must pay the costs.

For example, the owner may not be allowed to cut down trees on land within a certain distance of the bird's colony or to plow land to create a firebreak or even to farm. With all that power, the government is likely to be lavishly wasteful of some resources (such as land) while ignoring other ways to protect the species (such as building nest boxes for woodpeckers). When the land is essentially a free good for government officials, they will almost inevitably waste it.

The point of these two examples is that when people have to pay for what they use, they carefully weigh the costs and benefits. When they themselves don't have to pay, they may be willing to put heavy costs on others.

Although incentives are important, they are not the only factors in decisionmaking. For example, income levels affect how people deal with environmental problems.

People with high incomes tend to have greater concern about the protection of natural environments, such as old-growth timber or the habitat for rare plants or animals, than do people who have less income. Lower-income people may want to see those same lands managed to produce more food, raw materials, and jobs.

Very poor people, lacking the essentials of environmental protection such as drinking water free of parasites and of microbial diseases, may want the very basics of environmental quality. To some readers of this book, the desire for drinking water free of parasites may not even seem to be an environmental concern. But for people on the edge of poverty, safe water may be an environmental improvement. In sum, the same incentive may not have the same effect on people in different circumstances.

Other factors matter, too. Cultural norms and traditions affect how people value parts of their environment. Whether people toss litter on the ground or out of a car window reflects their education and probably the attitudes of those with whom they associate.

5. In market exchange, people can only gain at the expense of others—right?
Wrong!
Voluntary exchange—that is, market trading—creates wealth.

Counterintuitively, simple voluntary exchange can create wealth. Both sides can gain. One way to understand this principle is to think about something that people really disagree about—say, music. John likes opera. Jane likes rock music. If John has a rock concert ticket and Jane an opera ticket, just exchanging the tickets will make each person better off.

Trade can create value in three ways:

1. *Trade channels resources, products, and services from those who value them less to those who value them more.* Without any change in production, the trade of the opera ticket for the rock concert ticket produces value. Keep in mind that value can be subjective.

2. *Trade enables individuals to direct their resources to the activities in which they produce the greatest value so that they can then trade the fruits of those activities for the items they want for themselves.* The farmer in central Montana who grows wheat produces far more than he wants to consume. He trades the wheat for income to buy coffee from Guatemala, shoes from Thailand, and oranges from Florida. Value can and does change on the margin; traders get rid of excesses in one area to get products that meet shortfalls elsewhere. The Montana farmer might have been able to grow oranges, but given the cold Montana climate, doing so would have squandered resources. Trade enables people to obtain many things they would not have the proper talent or resources to produce efficiently themselves.

3. *Trade enables everyone to gain from the division of labor and from economies of scale.* Only with trade can individuals specialize narrowly in computer programming, writing books, or playing professional golf—developing highly productive skills that would be impossible to obtain if each family had to produce everything for itself. Similarly, sales from large automobile factories that bring the cost of cars within reach of the average worker would not be feasible without large-scale trade that enables the product of one factory to be sold in a wide market area.

Resource owners gain by trading in three different ways: across uses (for example, out of low-valued crops into ones that earn more money), across space (marketing products across geographic distance to different states or nations where the resource is more valuable), and across time (such as gaining from conservation or speculation by saving resources until they become more valuable).

Such trades are occurring in the West with respect to fishing. In recent years, more people have sought high-quality "blue-ribbon" streams for fly-fishing. But some streams dry up in summer months because farmers, who have rights to divert stream water, use large amounts of it for their fields. To keep the streams full of water and the fish thriving, some fishers are willing to pay cash to lease the farmers' water rights. In those cases, they are shifting the water to a higher-valued use.

Such exchanges are increasing. Andrew Purkey, then the executive director of the fledgling Oregon Water Trust, probably didn't know that he was starting a revolution in water markets when he paid a rancher $6,000 to *not* grow hay one dry year in the early 1990s. The water the rancher would otherwise have used on his fields remained in the stream and supported the fish. Since then, other organizations have taken up the task. For example, the Scott River Water Trust in northern California leases stream water from farmers. Between 2008 and 2011, the number of documented Coho salmon in a stretch of the Scott River rose from 62 to 340, a fivefold increase.[6]

Other farmers might gain by selling some of their water rights to growing cities, which can then save the cost (and the environmental disturbance) of building another dam or a saltwater desalinization plant to make fresh water from ocean water. When the law allows such trades among willing buyers and willing sellers, both buyer and seller are made better off. Value is added to the water's use. Wealth is created.

Unfortunately, right now the federal government and many western states have laws that pose obstacles to trade in water. These obstacles, such as the rule that only some uses of water are allowed for trades, tend to keep the water in agriculture, thereby reducing efficient use and conservation.

Even trade in garbage can create wealth. Consider a city that disposes of garbage in a landfill. If the city is located in an area where underground water lies near the surface, disposing of garbage is dangerous, and very costly measures would have to be taken to protect the water from contamination. Such a city may gain by finding a trading partner with more suitable land where a properly constructed landfill does not threaten to pollute water. Such a landowner may be willing to accept garbage in return for pay. If so, both parties will be better off.

6. What do profits achieve for the environment? And for that matter, what do they do for consumers?

In a competitive market, profits and losses direct businesses toward activities that increase consumer wealth and conserve resources.

17

Profits attract producers and sellers. Where there is profit to be made, consumers benefit from the increased competition to produce the good or service. And profits lead to careful use of natural resources because a firm can increase profits by reducing its costs (so long as that doesn't reduce the value of its output).

Producers spend untold hours figuring out how to save on resources in search of increased profit. That's why steel high-rise buildings today require about one-third the amount of steel they needed several decades ago and why producers prefer a fiber-optic cable made from 60 pounds of sand that can carry 1,000 times more information than a cable made from 2,000 pounds of copper.[7]

A century before "recycling" became a household word in the United States, companies had figured out how to reduce costs by turning waste products into useful ones. While conducting research on waste, Pierre Desrochers, a Canadian geographer, found a description of recycling written by the famed British mathematician Charles Babbage in 1835. Desrochers writes:

> Babbage described how horns from livestock were used by many other industries early in the nineteenth century. Some were made into combs and a substitute for lantern glass; others were carved into knife handles and the tops of whips. The processing provided fat for soapmakers, glue to stiffen clothes, and fertilizer for farmers—even toys for children.[8]

Profits reward those who succeed in producing goods and services that people are willing to buy at a higher price than the cost of supplying them.

Losses have their place, too. They penalize those who have not been able to discover how to create more value than the cost of production. In effect, people are telling a money-losing firm that they want to see the firm's resources go to other products or services that are more valuable to them.

Large profits have a way of disappearing, however. The competition of new entrants, drawn by profits, gradually lowers the sales of existing firms and often their prices as well, thus reducing their profits. Entry continues until profits fall to what economists call "normal rates of return." Entry then stops. The first firm to innovate successfully may make above-normal profits, but the profit rate falls as competition heats up.

18

Usually, an entrepreneur seeking to exploit a new profit opportunity must (a) discover the new opportunity and (b) find investors willing to take the risk that profits will in fact be gained. It may also be necessary to persuade potential buyers of the value of the new product or service. All of those activities are costly.

But the expected profit provides an incentive to persevere for entrepreneurs, investors, and those who must sell the idea to investors and the product to buyers. It rewards them for making the investments of time, effort, and money needed to accomplish their tasks. New ideas may require years of effort before they reach fruition. Expected profit is the carrot to attract those efforts.

Which producers obtain the opportunity to use scarce resources? The answer is those investors and producers who expect more profit. They are willing to pay more, when necessary, to get their raw materials. Those who expect losses are not willing (and those already experiencing losses are less able) to compete for the same resources. Producers who take resources from high-valued to low-valued uses in a market setting sustain losses.

Moving resources to higher-valued uses confers profits. Losses discipline investors and producers whose expectations are too optimistic, whereas profits reward those who make the best use, as seen by consumers, of the available resources.

7. Information is the resource of the modern age; every decision should be made with full information. Right?
Wrong again!
Information is a valuable, but costly, resource.
Let's say that a private owner decides to build a landfill for garbage. The owner is liable for damages if waste deposited in the landfill leaks out and harms others. So the owner must decide how to prevent leaks and how to clean them up if they occur. Spending too little on preventing harm from escaping pollutants could bring costly lawsuits. But spending more than is necessary imposes needless costs and wastes resources. How many resources should be devoted to preventing harm? In other words, how much should be spent? That is the question facing the owner.

To make the decision, the owner must have good information. Where is the groundwater underneath this land? How effective will a clay cap be? What liner will be the safest? Yet gathering the information needed to make a better decision is costly.

19

This owner, operating in the private sector, has an incentive to gather just enough information—not too much and not too little—because both the costs and the benefits of seeking more information fall on the owner's shoulders. By weighing the costs and benefits of more information, the owner won't end up with perfect or complete information but will make a reasonable choice that is based on the expected costs and benefits of seeking more knowledge.

Now suppose that a government regulator (perhaps someone in the local zoning office) has the authority to decide whether the landfill can be built. This individual's desire for information will be very different. If damage occurs, the regulator could be blamed, so his or her incentive will be to require as much information as possible before allowing the landfill to be built. Furthermore, the regulator doesn't face the costs of seeking more information or the costs of choosing an expensive way to reduce risks from the landfill.

The regulator may ask for study after study to make sure that the proposed landfill will really be safe. Not surprisingly, people running small businesses often complain that regulators are simply asking for too much paperwork. And if the project later falls through, the regulator bears none of the associated costs.

In other words, the information-gathering process is affected by where the costs fall. A regulator might demand too much information, but under some conditions the owner might seek too little. Suppose the property rights of neighbors are not effectively protected under law, and the private owner of the waste site is not accountable for harm caused by materials escaping the site. In that case, the owner may neglect to pay the cost of preventing pollutants from seeping out of the site because the costs of any harm are likely to fall on others. The incentive to seek additional information is weak because the owner doesn't expect to pay the costs of making a poor decision.

Important decisions require good information. Should a forest be cut now and replanted? Should the owner of a potentially polluting hazardous waste site be forced to spend several million dollars in a cleanup effort? Should exploration for new mineral deposits be conducted now or later? Should an environmental rule be further tightened?

Each of these decisions involves gathering scarce and costly information, and each decision must be made without complete information. But the information-gathering process will be shaped by the incentives facing the decisionmaker.

8. New technology may be cheaper, but doesn't it destroy the environment? Wouldn't we be better off, environmentally, if only older, tried-and-true technologies are allowed?

No.

Advanced technologies typically help the environment because they decrease resource waste and increase resource productivity.

Sometimes we wish for the good old days before we suffered from the pollution and congestion caused by automobiles. But our ancestors didn't think of cars that way. To them, the advent of the automobile was a blessing because it meant that horses no longer clogged the streets with manure. An even bigger effect is that thousands, perhaps millions, of acres of land have reverted to forest because the land is no longer devoted to growing grass and hay for horses. Also, new farming technologies allow for more production from fewer acres, freeing still more land for reversion to habitat and recreation.

Yes, automobiles do pollute. But today's cars emit a tiny fraction of the pollution emitted by the cars of the early 1970s, before serious pollution control began. And although even very expensive and clean-running electric cars require energy from burning fuel in power plants, the emissions from such plants have gone down drastically, too, as owners have adopted low-sulfur coal and technical devices to reduce pollution. Advances in technology continue to make cars cleaner and safer, just as diesel train engines replaced dirty coal-fired locomotives, and gas and electricity replaced coal for home heating.

New technology is almost always adopted because it is more efficient. It generally uses fewer resources to produce the same result. Stifling new technology unnecessarily forces us to forgo additional gains that could be delivered over time.

9. If the rich countries would just stop consuming so much, couldn't we all live more comfortably on this planet?

No.

As people's incomes increase, their willingness to pay for protecting the environment increases.

Even poor communities are willing to make sacrifices for some basic components of environmental protection, such as access to safe and clean drinking water and sanitary handling of human and animal wastes. As income rises, citizens raise their environmental goals. Once basic demands for food, clothing, and shelter are met, people demand cleaner air, cleaner streams, more outdoor recreation, and the protection of wild lands. With higher incomes, citizens place higher priorities on environmental objectives.

Economists have frequently noted the connection of income with better environmental quality. One study, for example, showed that in countries where rising incomes reached about $8,000 to $10,600 per year in 2015 dollars and where there initially was an increase in certain types of air pollution, air pollution began to decline.[9] Also, the kinds of water and air pollution (water with parasites and indoor air pollution) that very poor people confront fall steadily with rising incomes.

Another study suggests that the willingness of citizens to spend and sacrifice for a better environment rises far faster than income itself increases—more than twice as fast, in fact.[10] The same willingness and ability to pay for a better environment declines with falling income.

One can learn from advertising for members of the Sierra Club. Readers of *Sierra* magazine (most of whom are club members) have an average income of $92,374[11] compared with $53,046[12] for the average American family. That preference is another indicator that wealthier people are especially concerned about environmental matters.

One implication of this link is that the wealthier the people of the world are, the more concerned about the environment they will be. Similarly, if income falls, people will be less interested in paying for environmental protection. Policymakers should also recognize that if improvement in environmental quality can be achieved at a lower cost—rather than wasted through bureaucratic red tape, for example— public support for additional environmental measures will be greater. Policies that do not deliver good environmental quality at the least cost to the economy reduce the citizens' willingness and ability to pay for environmental quality measures.

10. What is the single most common error in thinking about the economics of environmental policy?

The most common error in economics, as in ecology, is to ignore the secondary effects and long-term consequences of an action.

It is easy to overlook the unintended side effects of an action, especially if those effects will not be experienced soon. When individuals are not personally accountable for the full costs of their actions, they tend to ignore the secondary costs of what they do.

Consider the classic case of overgrazing on a commons, a pasture open to all herders for cattle grazing. Each herder captures the immediate and full benefits of grazing another cow but may hardly be aware of the reduction in next year's grass that the extra animal grazing this year is causing. The individual herder bears personally only a fraction of the costs—the reduced grazing available next year because of excessive grazing now—because all users share the future costs, whether or not their own cattle graze to the greatest extent they can. However, if the herders removed any of their own cows, they would bear the full burden of reducing their use. Thus, each herder has an incentive to add cattle, even though the pasture may be gradually deteriorating as a result.

This situation is known as the tragedy of the commons.

A similar problem can occur when a fishing territory is open to all fishers. Individual fishers capture all the benefits of their own harvesting of more fish now, but they pay only a small part of the total future costs of that action—the reduction of the fish population for future harvest. It is easy to ignore the indirect costs that will occur in the future, especially if the fisher will not ultimately pay the full, true cost of his or her actions.

Government decisionmaking provides additional examples. It is typical for cities to be years behind in the maintenance of their water delivery systems. The cost of a repair that will reduce water leaks is borne now, whereas much of the benefit lies in the future. The present costs tend to be more vividly seen and felt than the future benefits, so repairs are often postponed even though doing so makes the future costs much larger. The fact that the costs of postponement are not as direct or as immediate naturally encourages the costly postponement of maintenance.

Conclusion

Those 10 points provide a good start toward understanding how economics applies to environmental decisionmaking. The principles lay the foundation for understanding, first, how cooperation can help to protect the environment and, second, why conflict often occurs instead. Cooperation is the subject of Chapter 3.

3. Rights: How Property Rights and Markets Replace Conflict with Cooperation

Tom Bourland faced a challenge. As a wildlife biologist at International Paper Company in the mid-1980s, when it was one of the world's largest private forest owners, he wanted the forests to teem with wildlife—deer, bears, beavers, and woodpeckers. He didn't want his company simply to produce wood and pulp. Experimental efforts by the company proved that it was possible to enrich the forests in Louisiana and east Texas with wildlife, but doing so was costly. How could he make the improvement of wildlife habitat an integral part of the company's goals?

Bourland found a way. He knew that hunters, fishers, and campers loved the woods as well as he did and would pay for the opportunity to use it, but only if the woods were full of wildlife and diverse vegetation. So he began a program to market the recreational opportunities of that land. Once he began to make some money for International Paper this way, he had the clout within the firm that he needed to enhance the habitat. He could, for example, expand his use of prescribed burns to encourage new streams to keep the water clean so fish would thrive.[1]

Many people think of protecting the environment as a struggle characterized by noisy, bitter, and protracted political conflicts. In the Pacific Northwest, for example, communities have been divided for many years, as loggers and conservationists square off over how much land should be logged and how much should be set aside for endangered species. Yet Bourland's experience indicates that preservation does not always have to be a struggle.

Competition and conflict are inevitable in a world of scarcity, and they will occur whether markets or the government is the vehicle to achieve environmental protection. However, a market society can channel energy and enthusiasm into constructive action that often achieves both protection of the land and use of its resources, as Tom

Bourland did with International Paper's southeastern forests. (In 2006, the company sold most of its timberland in a restructuring—some of it to the Nature Conservancy and the Conservation Fund.)

Markets minimize political conflict because they depend on voluntary cooperation. Competitors increase their market power and wealth by providing what consumers want at low cost. When ownership is clear and wrongful harm is forbidden, competitors cannot use force to increase their power; instead they must cooperate with others for mutual gain.

But sometimes the ingredients necessary for the smooth channeling of competition into productive activities are missing. This chapter shows how privately held property rights and market exchange encourage cooperation and conservation, and also it shows what happens when the necessary ingredients are missing.

1. Property rights are a necessary condition for market exchange.

A market is a place—or in today's experience, a "space, physical or virtual"—where property rights (ownership) are traded. A property right is the right to use something; the right to exclude others from using it; and usually the right to sell, rent, or lease it. The person who purchases a farm, for example, is buying the owner's right to the exclusive use of that farm, the right to rent or lease the land to others, and the right to sell it later.

For a market to function properly, property rights must satisfy three conditions. This author calls them the "3-Ds." Property rights must be defined, defendable, and divestible. Markets can be effective only to the degree that property rights are 3-D.

Rights must be defined. Most goods and services that we deal with every day are well defined. In fact, a lot of effort goes into making sure exactly what they include. Before purchase, land is often surveyed, and the boundaries of property are recorded in a local government office so that any dispute about where one person's property ends and another's begins can be easily settled.

The rights that go with owning land vary. Is burning wood in a fireplace permitted? What about building whatever you want on the property? What is considered a nuisance? Is there a homeowners' association that you must contract with to buy the property? A potential buyer would be unwilling to spend much money on a new home if ownership and ownership rights were not clearly defined.

26

Sometimes, however, ownership isn't defined at all. The water flowing in most streams in the eastern United States has no owner, although the owners of property next to the water have a right to reasonable use of the water. No one really owns wildlife either, although state governments have some level of control over it.

And sometimes ownership, although defined in theory, is not clearly defined in practice. People may be unsure about the precise boundaries of their properties. A fence on the boundary of two pieces of property might be poorly maintained if neither landowner is sure who owns it. Uncertainty about the definition of property rights and who is responsible for the upkeep of a property reduces its value.

In recent years, environmental regulations have sometimes made the definition of land ownership unclear. For example, regulatory bodies such as the California Coastal Commission have required some landholders to provide public access to their lands. And under the Endangered Species Act, the federal government has authority to protect habitat by limiting the way that land is used.

Such situations make the effective ownership unclear, especially to owners of property that is nearby or in similar situations. Uncertainty about ownership lowers the value of property. If the rights to a resource or a good are not defined or are poorly defined, the value of that resource falls. Potential buyers will be unwilling to pay much for the ownership of rights that aren't clear.

Rights must be defendable. Usually in the United States, property rights are readily defended. The courts will back an owner's property rights. However, if for any reason rights to a resource are difficult to defend against theft, harm, or trespass, the resource's value to the owner falls.

A farmer's land that is too easily accessible to others can be robbed of its produce. The smell from a newly established hog-feeding operation may invade another owner's property. Land may become contaminated by hazardous waste from an unknown source. If the owner is unable to fully defend his or her ownership rights, the property is worth less than it would be otherwise.

The government can take away some elements of property rights through excessive regulation. For example, when the government designates land as a wetland, the U.S. Army Corps of Engineers can demand that landowners either not disturb it or pay the costs of adding wetlands elsewhere.

27

Regulations to stop actions that would harm others (causing a flood, for example) are a legitimate use of the government's police power. However, regulations that instead require a person to provide a public service (such as protecting a wetland without payment) may violate his or her property rights. The courts have been addressing these issues in recent years and, in some cases, have restricted the government's ability to make such regulatory demands without compensation. The David Lucas case mentioned in Chapter 1 is an example.

Rights must be divestible. If property rights to a resource cannot be divested, meaning that the owner is not free to sell or lease the resource at will, those rights are less valuable to the owner. That component of the resource's productivity is not as likely to be well preserved and well used.

In normal day-to-day living, we can usually sell what we own. Craigslist, eBay, and most newspapers are filled with advertisements for property being offered by its owners—everything from used furniture to automobiles and boats. Most people in the free world take for granted the ability to buy and sell property.

But there are important exceptions to some owners' ability to divest their land or resources.

In the recent past, people often saved on inheritance taxes by providing property to a son or daughter but holding it in trust for the grandchildren. The son or daughter could only obtain the income from the property, not the full value of selling it. This tactic tended to distort decisions about how to use and care for the property. The middle generation had an incentive to increase its income by overusing the property, thus decreasing the property's long-term value.

On most American Indian reservations, federal laws have led to a situation in which ownership of certain land that is passed from one generation to another must be divided among an increasing number of heirs. Reaching agreement among the heirs on what to do with the property becomes difficult, thus limiting the ability of the owners to properly manage or transfer the property.

The ability to divest property has enormous—but often unrecognized—effects, as a *Wall Street Journal* cartoon illustrates. A husband and wife are walking out of a home. The man says to the woman, "Their house looks so nice. They must be getting ready to sell it." The motivation to obtain maximum value from a potential buyer encourages people to maintain and improve their property. The ability to divest encourages stewardship.

2. Private ownership and protection of property rights provide each resource owner with both the means and the incentive to protect and conserve the resource.

Very simply, property rights hold people accountable. When people treat property negligently or carelessly, its value decreases. When they treat it with care, its value typically increases. Aristotle recognized this point more than 2,000 years ago when he said, "What is common to many is taken least care of, for all men have greater regard for what is their own than for what they possess in common with others."[2]

To be effective, property rights must be protected by law. In a society that protects property rights, people's resources, including themselves, are protected from harm. Such harm includes not only theft and assault but also pollution.

Protection against pollution occurs through the courts. In the United States, Canada, and other nations having legal roots in Great Britain, the courts have for centuries provided a way to stop individuals from injuring others by polluting.

When a pollution victim shows that harm has been done or that serious harm is threatened, courts can force compensation to the victim or issue an injunction to stop the polluting activity. Such recourse is through common law. "Common law" refers to the body of legal rules and traditions that have been developed over time through court decisions. Each decision helps to settle the details of the law by putting everyone on notice of what is expected, thereby reducing uncertainty and thus the need for future legal action.

It is easy to find examples of common-law protection against pollution, even going back more than 100 years. In the late 19th century, the Carmichael family owned a 45-acre farm in Texas, with a stream running through it that bordered the state of Arkansas. The city of Texarkana, Arkansas, built a sewerage system that deposited sewage in the river in front of the Carmichaels' home. The Carmichaels sued the city in federal court on the grounds that they and their livestock could no longer use the river and possibly were being exposed to disease. (They were able to sue in federal court because two states were involved.)

The court awarded damages to the Carmichaels and granted an injunction against the city, forcing it to stop the harmful dumping.

Even though the city of Texarkana was operating properly under state law in building a sewerage system, it could not foul the water used by the Carmichaels. Indeed, the judge noted, "I have failed to find a single well-considered case where the American courts have not granted relief under circumstances such as are alleged in this bill against the city."[3]

Another example of the protection of natural resources through the protection of property rights can be found in England and Scotland. There, in contrast to the United States, fishing rights along the banks of streams are privately owned, usually by landowners along the streams. These rights to fish can be sold or leased, even though the water itself is not privately owned.

Owners of fishing rights can take polluters of streams to court if the pollution harms their fishing rights. Indeed, after an association of British anglers won a celebrated case in the early 1950s against a government-owned utility and a private firm, it has rarely been necessary to go to court to stop pollution that damages fishing. Once established by precedent, such rights seldom need to be defended in court unless, in a particular case, the circumstances are new and unlike those in previous cases. When the courts are doing their job in protecting property rights, natural resources are protected more efficiently than through extensive bureaucratic controls such as contemporary environmental regulations.

The tradition that protected the Carmichaels in the 19th century still protects citizens today. However, in many cases, these common-law rules have been superseded by government regulations, often with worse results. For example, although the Carmichaels sued a city in a different state and won, the city of Milwaukee in 1972 tried to sue the state of Illinois for polluting its water. But the passage of the Clean Water Act in 1972 led a judge to dismiss the case because water pollution is now in the hands of federal agencies.

3. Market trades and market prices bring narrow personal interest into harmony with the general welfare.

Adam Smith recognized that voluntary exchange channels individual desire, as if by an "invisible hand," into socially beneficial activities. An individual may or may not care deeply about the happiness of others in the trade, but it pays that individual to act *as if* the happiness of others matters. After all, the more the seller can please the buyer, the more willing the buyer is to pay what the seller asks. Similarly, if buyers can please suppliers by offering to pay more, the suppliers will be more responsive to their desires.

Buyers and sellers may act with little knowledge of what any other person wants or needs. But market prices direct each person to satisfy the needs of others. Prices encourage producers to provide what members of society want most, relative to its cost, and to satisfy any particular want in the least costly way.

Consumers, too, are strongly influenced by market prices, and they, too, act as if they care about their fellow consumers. When prices increase, they consume less; when prices decrease, they consume more. By economizing when goods are scarce, they allow more for other consumers. In contrast, they purchase more when the goods are plentiful and there is a lot to go around. Actual and expected offers in the marketplace are guidance from the so-called invisible hand.

Consider energy markets. Each consumer of electricity chooses whether to use electric heat and how high or low to set the thermostat. In addition to demonstrating preferences about temperature, these decisions reflect the price of electricity. When prices are high, people will economize, thereby making more of the resource available for others. When prices are low, they will consume more.

These decisions, in turn, influence the decisions of others, even those made by other industries. For instance, individual consumer choices about electricity consumption affect how much aluminum will be produced and which producers will supply more than others.

Why? Consider aluminum production, which requires large quantities of electricity. Higher electricity prices raise the price of aluminum compared with substitute metals and especially raise costs for producers that use a lot of electricity per ton of aluminum made. As electricity prices rise, consumers—even if they have little or no knowledge of why prices are rising—will begin to use electricity more carefully. For example, builders of new homes may choose to rely more on natural gas and less on electricity. These choices gradually move sales away from the expensive energy sources and toward conservation or substitute energy sources. These forces tend to reduce electricity prices. Thus, aluminum producers don't have to pay as much for energy, and the prices of their products can moderate, too.

Similarly, aluminum producers who conserve on the use of electricity enjoy a competitive advantage and are likely to produce a larger share of aluminum sold in the market.

4. Private rights and market exchange minimize conflict.

The conflicts over environmental resources that drag on and on are almost always political conflicts. Government decisions favor the side with the most political power (that is, the one with the greatest ability to influence elected officials and regulators). The losing side must accept the decision and usually pay taxes for a result it does not like. The political decisionmaking process is often a zero-sum game. In other words, what one person or interest group wins, another person or interest group must give up without any prospect of gains from trade.

In contrast, market exchanges tend to be win–win. Even though there is plenty of negotiation and disagreement in the marketplace, the solutions that people agree on are ones that both parties want—at least compared with available alternatives. And a would-be buyer whose offer is rejected does not have to pay. Because market decisions are voluntary, people will not agree to an exchange unless they think it will improve their situations.

Often individuals or organizations make different decisions in the political arena than they would in the free market. This point can be illustrated by an experience of the National Audubon Society, which we met in Chapter 1.

Audubon owns the Paul J. Rainey Wildlife Sanctuary in Louisiana, a wildlife refuge that provides nesting grounds for snowy egrets and other rare birds. Audubon allowed drilling for natural gas and oil there from the 1940s until 1999.

When the potential of its energy reserves (mainly natural gas) became known, the society chose to exploit its deposits. Audubon experts and the biologists for the oil companies worked out methods of drilling and production that would not harm the snowy egrets and other birds and animals. The companies had to meet Audubon's strict stipulations; they were not allowed to drill during nesting season, for example. Audubon gave up substantial income by demanding the stipulations, but by doing so, it continued to protect the natural habitat.

The cooperation between the National Audubon Society and the production companies benefited both. Revenues for Audubon totaled more than $25 million over the years, helping it pursue its mission, both at Rainey and elsewhere. The producers of natural gas earned revenues that would have been lost if Rainey had been closed to them.[4]

In 2010, the society again considered opening the sanctuary to oil and gas drilling. Hurricane Rita had caused damage to the sanctuary, and Audubon officials needed funds to restore it. So far, however, no agreement has been reached.[5]

Such cooperation is missing on land and water owned by the government in and around Alaska. "We're using every weapon in our arsenal to avert this desecration of our shared natural heritage," wrote David Yarnold, president of the Audubon Society, in 2012. "The battle to protect the Arctic is a touchstone for conservation and the future of wilderness and wildlife on our planet."[6]

The fact that the government, not Audubon, owns the waters means that the actual stipulations for exploration and drilling would result from a political process. Audubon might have an effect but not the control that it has over its own preserve, nor would it receive any benefit from the drilling. So it is understandable that Audubon opposes drilling and instead adopts an all-or-nothing vocabulary in which the land is sacred and prices are never discussed.

As the Rainey example shows, ownership fosters cooperation. Such cooperation is important not only when something like energy development conflicts with environmental values but also when environmental values themselves conflict.

Here, too, markets offer tools for cooperation. For example, how should pristine open space around a backwoods lake be used? It could be open to people from nearby cities for hiking, camping, fishing, and swimming. Or it could be closed to all visitors (except, perhaps, research biologists) to protect its rare flora and fauna. Both alternatives have merit. Some environmentalists will want one option; others, a different one. In such cases, it's hard to know what is really in the public interest.

We often witness acrimonious discussions at public meetings, read angry letters to newspaper editors, and learn of the pressures on government officials when environmental decisions are made politically. Former allies, all viewing themselves as environmentalists, take different sides depending on their own goals and the expected costs and benefits (to their own narrow goals).

If the decision is made privately in the market, however, there will be much less acrimony, and discussion will be more productive. Suppose the owner of the land around the lake wants to sell it. Someone who wants it for a campground will have to determine whether the public will support a campground by paying fees. Those who want to set it aside as a wildlife preserve must determine whether they or their associates can generate the funds to purchase it. People who get involved in the bargaining have an incentive to find ways to meet their own goals at the least cost to all. If they can come up with a solution that creates more value, they can ask more in return.

That is why each party has a good reason to consider additional uses that may be compatible. Although each party has the right to be dogmatic, favoring only specific narrow goals, such an attitude may prevent possible opportunities to achieve many of that party's goals. The result of the negotiation could be a campground or a private nature preserve, or an innovation that allows for both.

Not everyone will get what he or she wants. Those who are not willing to provide any resources will probably be outside of the negotiations; only those who have something to offer are involved. Political decisions do not please everyone either. The key difference is that in a private setting, those who do not engage in the negotiations or whose offers are rejected do not have to pay for the outcome. In contrast, when a decision such as establishing a park is made publicly, taxpayers usually bear the costs, even those who had no say in the decision and who may not even be able to use the park.

5. Market prices provide knowledge that is complex, dispersed, and constantly updated.

Market prices provide participants with information from all corners of the market, which increasingly means from all over the world. The information is in a highly condensed form. The reasons for a price rise or a price decline are not conveyed—just the vital fact of increasing or decreasing scarcity compared with other goods and services. Prices do, however, carry something else—a powerful incentive for buyers and sellers to act on the information.

Furthermore, market prices adjust constantly to all of the factors that affect supply and demand, providing each buyer and each seller with up-to-date information on changes in relative values in the world around them. Without market signals, it would be nearly impossible to evaluate the effect of (or even keep track of) all these bits of knowledge regarding scarcity in many uses and locations. Yet each one is relevant to the cost and value of what is preserved, produced, and offered in the marketplace.

The value and price of logging rights in southwestern Montana, for example, depend on timber supply and lumber demands all over the globe. The same is true for the price of wheat in Kansas and the value of hunting on safari (with a gun or a camera) in South Africa. Prices change constantly, telling everyone about changing factors around the globe.

When resources are not privately owned and traded in open markets, however, this vital flow of information is missing. That is the case with our national parks.

Most of the funds for the national parks are tax dollars appropriated by Congress. Park visitors pay only a small fraction of the cost of the services they receive—a little over 8 percent of the cost of park operations in 2014.

Thus, what visitors choose to pay for gives park managers little information about how much the various services are worth to visitors. To learn what people want, the National Park Service has to rely on expensive surveys and polls, which can reach only a small number of people.

Without knowing what visitors want, park managers allocate their budgets with critical information missing. Should the roads, buildings, and sewers be improved? Should more rangers be hired for interpretive programs? Should campgrounds be kept open and operating hours lengthened? Should new campgrounds be added? Park managers cannot know the answers because a market is lacking. Hence, they may choose what they prefer, not what visitors want.

Even when park managers have valid information, they may not have an incentive to use it. For example, although visitors pay only a small part of total park operating expenses, they do cover many of the costs of a park's campgrounds. But who actually receives the money can affect how it is spent.

In the summer of 1996, the managers of Yellowstone National Park, in an attempt to save money, closed a campground. In fact, this campground was profitable—that is, it earned more than it cost to operate. But the revenues from the campground went to the U.S. Treasury Department in Washington, D.C., not to the managers of Yellowstone National Park. By closing the campground, the park managers reduced their expenditures, but they reduced the revenues to the U.S. Treasury Department by a larger amount. Keeping the campground open would have drained their park budget, even though the campground was worth more to the users than its operating costs.

In contrast, for owners of private campgrounds, amusement parks, museums, and other attractions that also draw visitors, information is always flowing, and managers always have an incentive to respond to that information. These owners must pay for the resources they use and collect the needed revenue from customers. If people don't come to the museum, revenues fall. Owners must do something to attract customers who are willing to pay the full cost (or donors who will pay for them) or they will have to close their doors.

Most states fund their parks with state appropriations, but state park visitors on average pay 39 percent of the total operating cost—a bigger percentage than is raised by the National Park Service. A number of states have found that when their park systems rely more on the visitors than on the state legislature for revenues, the parks begin to provide services that visitors want and are willing to pay for. One state, New Hampshire, gets its entire operating budget from user fees.

These factors have led one expert, economist Holly Fretwell, to recommend that national parks be "franchised" or "outsourced" so that their staffs will be more responsive to their customers. Kurt Repanshek, editor of *National Parks Traveler*, commented on the idea: "Already the Park Service contracts with others to manage its lodgings, restaurants, and many campgrounds, and it relies heavily on volunteers to cope with visitors. So why not go all in? Would it make a stronger, more efficient, and better managed park system if individual units were treated, say, as franchises that were independently managed?"[7]

Although Repanshek didn't necessarily support the proposal, he found it intriguing.

6. Markets encourage solutions that are appropriate for specific circumstances.

The market system spurs conservation. Producers benefit when they save on resources because their costs decline. They also benefit, as do their customers, by developing new technologies that increase the value of the output from the resources they use.

But proper resource conservation differs from one place to another and from one time to another. An excellent solution to a problem in one situation may not work well for a seemingly similar problem in a different setting.

Consider the question of cloth versus disposable diapers. Although cloth diapers have the advantage of being used again and again, they must be washed and dried each time. Such laundering requires hot water and detergent, disposal of wastewater, and perhaps heat to dry the diapers. In addition, if a diaper service is used, transportation to and from the customer also places a demand on the environment.

Disposable diapers are used only once. Each new diaper is produced in a manufacturing process that requires new cellulose fiber and plastic liners, and disposal requires transportation and space in a landfill.

In a place where water and the energy to heat it are scarce but landfill space is plentiful—the rural West, for example—disposable diapers may be better for the environment. But where landfill space is scarcer and more expensive but water is abundant—urban areas in the East, for example—cloth diapers may be environmentally better.

If those who make and those who buy diapers pay the full cost of their use, market prices will automatically signal the relative resource costs for the specific situation. Higher water prices signal scarce supply that makes water conservation more valuable. This situation encourages the use of disposable diapers. The price signal not only gives information, but also encourages users to choose the product that, under their specific circumstances, places less total cost, including environmental costs, on society. Thus, the market system rewards conservation of the more highly valued resources.[8]

However, the price signals work only if producers, consumers, and third parties are secure against polluters. If nonconsumers suffer air pollution from diaper-service trucks that travel to and from their customers' homes, for example, then the consumers are not paying the full cost of cloth diapers, and the market signals to the producer are distorted. Similarly, if landfills are too cheap (that is, if the full cost of maintaining them is not being paid by those who use them), consumers may be getting an incorrect market signal. They will tend to use the landfill more than is optimum because they are not paying its full cost.

7. Private ownership provides freedom and a powerful incentive to innovate.

Over the past century, new technology has led to less pollution and to the use of fewer raw materials per unit of output. This has been true for steel mills (once fiery behemoths belching smoke but now relatively clean, with many using scrap steel as their raw materials) as well as for aluminum cans (which over time have been engineered to become thinner and thinner). Similarly, new technology has reduced the amount of energy required to produce a specific amount of output. As Figure 3.1 indicates, the amount of energy per unit of gross domestic product has been falling. Innovation can benefit not only manufacturers and users, but also the environment.

Figure 3.1
U.S. ENERGY INTENSITY PROJECTED TO CONTINUE ITS STEADY DECLINE THROUGH 2040

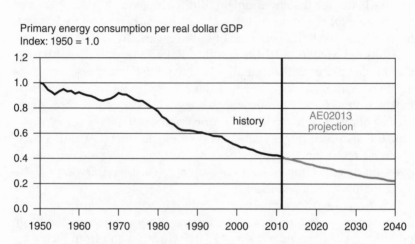

Primary energy consumption per real dollar GDP
Index: 1950 = 1.0

SOURCES: U.S. Energy Information Administration, *Annual Energy Review*; and *Annual Energy Outlook 2013*, http://www.eia.gov/todayinenergy/detail.cfm?id=10191.

NOTE: Primary (raw fuels) energy consumption per real dollar of GDP (Index: 1950 = 1.0).

Innovation is essential to progress, but it means change, and change is always difficult. Usually, continuing to do something the way it has always been done is easier. A market society encourages innovation but also keeps it within limits and tests whether it is getting good results.

To have an incentive to innovate, an entrepreneur must be able to benefit personally from a change. This incentive comes through ownership. Owners and investors can save money by using new techniques that reduce the use of resources—and can earn more revenues with new or better products. At the same time, they don't want to lose all of their wealth in a reckless pursuit of profits, so they are cautious about pursuing zany ideas and will abandon those that prove unworkable.

In a private system, the individual's wealth—not society's—is at stake. Other people do not pay for innovations that go awry.

Innovation is so embedded in a market system that it may be hard to imagine a system without it. However, the socialist economies of Eastern Europe were, by and large, such a system. The Trabant automobile, produced in East Germany between 1957 and 1991, illustrates what happens without innovation.

In those years, central planners ran East Germany, and private property rights were largely missing from the economic scene. Managers had little incentive to innovate and little freedom to act at all. Production plans were dictated by central planners, and managers were mainly rewarded for meeting quantitative output goals.

It wasn't until after the Berlin Wall came down in 1989 that many Americans saw the Trabant. An American auto magazine, *Car and Driver*, brought the Trabant over to have a look at it. On the positive side, the editors reported that the car provided basic transportation and was easy to fix (similar things were said about the Ford Model T in the early 20th century). But its top speed was 66 miles per hour, it was noisy, and, the editors said, it had "no discernible handling." It spewed "a plume of oil and gray exhaust smoke" and didn't have a gas gauge. In fact, the Trabant's exhaust was so noxious that the Environmental Protection Agency refused to let the *Car and Driver* staff drive it on public streets.[9]

The Trabant was backward, dirty, and inefficient because its design was the same as when it was first manufactured in 1957. The last model had been introduced in 1964, and because market pressures were absent, there had been no technological change since then. In contrast, during the same years, the Volkswagen, the "people's car" of West Germany, was continually updated. By 1989, Volkswagen vehicles were efficient and caused little pollution. Pollution control laws were a factor in reducing emissions, but fuel economy and performance alone would have brought some of the reduction. The safety, comfort, performance, and pollution control of the VW changed constantly for the better, whereas the Trabant stagnated.

8. Markets foster a variety of plans and actions, allowing unusual creative ideas to be tested.

In a market system, many mistakes are made. Entrepreneurs come up with new products, and customers reject some of them. Entrepreneurs try to save money by innovating, and some of the innovations don't work. About half of new start-ups don't last more than five years, reports Scott Shane of *Small Business Trends*, using government figures.[10] But the ideas that do work, the products that do sell, the businesses that do succeed provide the change that over the years transforms the economy and increases society's wealth by lowering costs for producers. In a world of economic freedom, those profits are diminished as market price falls, over time, toward the new, lower cost.

Change occurs rapidly in a market system because individuals don't need consensus or majority approval to pursue their ideas, as they would if their ideas had to be adopted by a democratic political process. As long as people don't harm others (except by decreasing the profits of high-cost producers), they can test their innovations, and in a market, "the early bird gets the worm." Successful innovators earn temporary profits; others must later adopt the innovations that work just to survive in the business.

History is replete with examples of people who have challenged the conventional wisdom. In the 1970s, it looked as though computers would be increasing in size and complexity, but a few hobbyists had a different idea. Some innovators put together a crude computer and began selling it as an assemble-it-yourself kit through *Popular Science* magazine. They created the first personal computer, revolutionizing the future of computers and changing the way people conduct business and leisure activities.

Such innovations occur in the environmental realm as well, often long before public policymakers recognize the need for change. As told in the first chapter, the story of Rosalie Edge and the Hawk Mountain Sanctuary reflects the advances made by people who think differently and innovate. Edge's view that hawks have an important place in nature is now conventional wisdom, and it is easy to forget that this idea was radical in the middle of the 20th century. Only because Hawk Mountain was privately owned could Edge exercise her vision of wildlife protection.

Although the National Audubon Society didn't protect hawks, it too was a pioneer in saving birds and bird species. Rather than relying solely on government preserves, it accepted both donations of natural areas that would become wildlife preserves and funding to manage them. Today, Audubon has 100 wildlife sanctuaries and nature centers around the country.

As discussed in the first chapter, Hank Fischer almost single-handedly brought back the wolf to Montana and Idaho through innovative techniques that reflected his understanding of incentives.

More recently, the Delta Waterfowl Foundation has paid farmers in the United States and Canada to protect nesting areas for ducks and geese on their farmlands. The program, called Alternative Land Use Services (ALUS) in Canada, enlists farmers in making their land more hospitable to waterfowl.

9. Private ownership makes resource owners accountable to the future.

People sometimes assume that a private owner has little incentive to protect a resource for the future and may be quite willing to destroy its long-term value for short-term gain. Only the government can truly preserve a natural resource, this line of reasoning goes, because the government, unlike the private sector, plans for the long run.

This common assumption, however, is largely false. Here's why.

The price of property reflects the future benefits that the owner expects to receive from owning that property. An office building, for example, is worth much more than it can earn in a few years; it is worth the estimated total rents, deducting the costs and accounting for the fact that a dollar received today is worth more than a dollar in the future. In economists' language, today's price of any asset is the capitalized value of the future stream of benefits (after removing the costs required to protect or produce those benefits).

Just as prices convey information about changing demand and supply all over the globe, a capital market—the buying and selling of capital assets such as land, buildings, bonds, or even corporate stock—provides information about the future as buyers speak today through their current market bids for ownership.

The operation of the capital market sends price signals about the use of assets such as land. Those few people who first see that a resource will rise in value can profit by buying and preserving it—and selling at a higher price when others recognize its value. Even a shortsighted owner who is personally concerned only with the present will respond to these signals because they change the current value of his or her assets. Of course, owners can ignore the price signals, but then they must deal with the resulting reduction in wealth.

In a sense, the value of a resource such as land is a hostage that ensures protection and good management by the owner. The desire to protect or enhance the value gives the owner an incentive to maintain the land's productivity and, where profits seem possible from adding value, to make investments that improve it. If the land is damaged, its value declines whether the damage occurs through misuse, negligence, trespass, or pollution. If necessary, an owner has the incentive to go to court against trespassers or polluters to protect the value of the property.

The incentive to look to the future is clear for conventional sources of income such as agricultural crops or housing developments. But it also holds true for goods of an environmental nature.

Consider television magnate Ted Turner, founder of the first 24-hour news network, CNN. After a successful and innovative career, Turner began buying ranches in the West and Southwest. On the Flying D Ranch, south of Bozeman, Montana, for example, he decided not to raise traditional livestock but instead to manage the ranch largely for bison and elk. To do this, he decided to increase the number of trophy animals over time, thereby increasing revenue and the value of the ranch.[11] Making this venture possible are the fees he charges hunters—$15,250 per elk, although there are also hunts that cost much less. Turner manages the ranch in a way that is pleasing to hunters and encourages the proliferation of diverse wildlife, not just elk, deer, and bison.

10. It is costly to establish and maintain private rights.

Private property rights benefit both the individual and society, but not everything is privately owned. One reason is that sometimes defining and defending property rights can be extremely expensive. Thus, at times a resource isn't valuable enough for anyone to make the effort.

Looking back over American history, for example, we notice that although the first colonists in New England established settlements where they lived, they made no effort to own the surrounding wilderness. It was full of wild beasts and inhabited by sometimes hostile Indians. It offered the colonists little or nothing that was useful for them.

Over time, people's view of wilderness changed. As settlements became more crowded, wilderness became more attractive. Settlers also had more tools than before (better guns and better saws, for example), so they could hunt, log, and live more safely in the forests. They began to establish ownership.

Today, of course, many people have a very different view of wilderness. For them, it is something to be prized—a place to get away from urban crowds and to commune with nature. That is why many Americans have encouraged the federal government to set aside large areas as wilderness and why individual Americans frequently try to buy a "little piece of wilderness" as a retirement or summer home. Over time, as wealth increased (and ecological effects became known), much American wilderness changed from being a worthless danger to a valuable resource.

Similarly, in the American West, most of the prairies and grasslands were so vast that no one—either American Indians or the pioneers who came later—thought about owning them. After the settlers arrived, they let cattle roam where buffalo had once wandered in giant herds. At first, so few cattle were roaming the vast rangeland that they caused no problem.

But that changed. Cattle became numerous enough that owners needed to separate one person's herd from another's. Had they been in the East, cattle owners would have used wooden fences, but in much of the West that was impossibly costly. There was less wood to build fences with, and the areas that had to be fenced stretched over square miles, not mere acres.

As the value of separating the herds increased, ranchers began to experiment. First, they organized crews of cowboys to round up the herds, and they began creating distinctive brands for their cattle.

Then some entrepreneurs invented barbed wire. When used to connect wooden posts, barbed wire is effective in keeping in cattle and, compared with wood fences, it is cheap. The invention of barbed wire reduced the cost of establishing property rights to herds of livestock and to the lands they grazed. Barbed wire allowed for the effective defense of property rights, a defense that had been missing.

This example shows that when a resource is not valued highly enough to justify the cost of establishing and defending property rights, the resource may not be privately owned. Over time, however, this situation can change. The value can increase, and the cost of protecting rights may go down.

The value of keeping water in streams and rivers, for example, has increased. In the second half of the 19th century when western water law developed, the chief demand for water in the West was to divert it—take it out of streams—for irrigation, mining, or other purposes. But in recent years, many people have wanted water left in the streams to protect the fish and to allow recreational use of the streams.

As the value of instream water has risen, environmental groups and fishing associations have pursued legal changes that allow them to privately define, defend, and transfer the right to use of those waters. The law has begun to allow instream flows to be treated much like private property, just as irrigators have for many decades been able to treat their rights to divert water from the streams as private property. One result has been conservation, such as the successful effort of the Scott River Water Trust in northern California, mentioned in Chapter 2, which compensates farmers for keeping water in the Scott River.

Conclusion

This chapter has shown that when property rights are defined, defendable, and divestible, markets turn the conflict that is caused by scarcity into a cooperative search for mutual benefits by buyers and sellers. However, property rights are not always 3-D. When they are not, environmental problems can arise.

4. Coercion: Protecting the Environment with Government Action

Newspaper headlines are full of conflicts over environmental issues. Should the government allow exploration for oil off the shores of southern states such as North Carolina and Florida? Should power plants be required to limit their emissions of carbon dioxide? Should the pesticide DDT be used to attack malaria-bearing mosquitoes in foreign countries?

In these cases, as in many others, the cooperative spirit illustrated by market transactions in Chapter 3 is missing. It has been replaced by angry, stubborn, and even extreme positions by both sides. The coercive power of government contributes to the conflict.

Yes, the government has a critical role in protecting the environment. The problem is that often the government intervenes in ways inappropriate to its critical role.

The government has a legal right to use force. Thus, on the positive side, it can police the protection of rights and prevent violence or fraud by one person or group against another, including rights against harm from pollution.

In addition to having these police powers, the government protects rights by recording claims such as records of ownership and sales of land and water rights. These activities help markets function better by making property rights clear. For example, such records help to more easily identify the owner of property that is the source of harmful activity. As noted in Chapter 3, property rights, when properly defined and defended, can further environmental goals.

Other ways that governments intervene, however, are not consistently beneficial. The government can own and manage resources such as land, wildlife, and water. It can also control how people and companies use resources, replacing market decisions with political decisions, as it does frequently with environmental issues.

Political decisions often lead to clashes because individuals in governments have different incentives from those who buy and sell in markets. Environmental statutes typically state specific goals and direct agency officials to use their authority to achieve them. In contrast, market participants must obtain cooperation voluntarily; they choose their own goals and priorities and rely on the government only for the rules and processes for settling disputes. These two types of systems provide different incentives.

The two settings also provide different kinds of information. Markets continuously provide both buyers and sellers with market signals, primarily in the form of prices. These signals inform buyers and sellers how they can do well by providing something that others want. The signals are frequent and finely tuned, in dollars and cents.

Government officials also obtain signals of a kind. In a democracy, the signals come from the voters in elections. However, unlike markets, the signals from the voters are infrequent and seldom specific. Exactly what mandate does a newly elected official have on a specific environmental question? Unless the answer to that question was the major factor causing voters to distinguish between candidates, the election outcome does not reveal exactly what voters want. Most voting is by candidate, not by issue.

This chapter explains how the government obtains information, what incentives its decisionmakers face, and in what predictable ways the government acts when it seeks to control resource use and environmental quality. This information illustrates the strengths and weaknesses of government and helps explain the results, both environmental and social, that we can expect from the political process.

1. Government plays a critical role in protecting individuals' rights to hold and use their properties and to be free from harms caused by others.

Governments give formal recognition to property rights, which are traded in markets. Markets are everywhere. From stock exchanges, in which billions of dollars' worth of ownership interest in capital are traded daily, to the farmers' markets that appear each summer along country roads, trade is a fact of life that benefits in some way all who participate.

Governments facilitate these exchanges. Governments use force (and more often the threat of force) to prevent theft and fraud. When people are confident that what they own is not going to be taken from them, they are more willing to buy and sell and to produce goods in the first place.

The economist Hernando de Soto discovered the critical role of protection of property rights when he studied the informal economy in Peru. He found that the Peruvian government, through neglect, bureaucratic inertia, and protection of privilege, had made it impossible for many Peruvians to open businesses.

Entrepreneurs had to go through labyrinthine approval processes that were costly, full of detailed requirements, and nearly impossible to complete. Many people in the poorer sectors had to operate their enterprises illegally if they were to have businesses at all. As a result, they did not have the basic protection of property rights that we generally expect government to provide. De Soto concluded that if society is to be cooperative and productive, property rights must be formally recognized so that people can plan for the future, knowing that they can keep what they earn and that any investment they make will not be taken away from them.[1]

Governments rarely *create* property rights. Although the history of property rights varies from place to place, property rights are usually established informally when land or other natural resources become valuable enough for individuals to work with them. Later, these informal rights are confirmed or codified by a government entity.

The discovery of gold in California in 1848 illustrates this process. The sudden increase in the value of land led briefly to conflicts among California miners. But soon the miners began to make agreements about how the land and the veins of gold would be divided. Claimants worked mines together, having made contracts that spelled out how finds would be allocated. They did so even though there was no effective government in those areas at the time. Later, when the federal government came west, it formalized the rights and provided legal protection.

Throughout most of the history of the United States, the government's role with respect to land and water was primarily to recognize, record, and protect individual property rights. Although the U.S. government claimed ownership to a large amount of land, most of it was gradually settled and became privately owned through various laws such as the Homestead Act of 1862. (This policy of divestiture or privatization ended late in the 19th century, when the federal government decided to keep many western lands. It still owns more than half of the land in four states.)

Once land was privately owned, state governments provided common-law courts—that is, the civil courts through which people decide disputes. Among those disputes were issues over damage from pollution, as discussed in Chapter 3. By enforcing property rights, government courts protected people from excessive pollution, just as they protected individuals from theft and from personal assault.

Protection against polluters depends on the plaintiffs' ability to show that the harm is occurring or is imminent. Succeeding in a civil case of this type is easier than gaining a conviction against a criminal offense, which requires proof beyond a reasonable doubt, but even so, proving the case can be difficult.

In some cases, there may not be sufficient evidence that the pollution was caused by the person or entity charged. Or the extent of harm may not be clear. Often there is a lag between the pollution and the harm, as when groundwater contamination shows up long after a leak has occurred. And often there is much scientific uncertainty about the effects on human health of differing levels of pollution.

So although Americans have the right not to be seriously harmed by polluters, the knowledge needed to demonstrate the harm effectively may be lacking. In addition, some important environmental concerns don't lend themselves to private suits at all. Although the residents downwind from industrial polluters such as the ASARCO smelters in Ruston, Washington, and Helena, Montana, were successful in suing the firm for harm from sulfur dioxide pollution, that form of recourse is not manageable for individuals who suffer from smog in Los Angeles or Houston. Smog in cities is caused by millions of scattered polluters, especially cars, and there are millions of scattered smog victims.

To act under common law, there needs to be a single harmed individual, or relatively few such individuals, who have the ability and the incentive to protect themselves through the courts. Similarly, it is important that specific identifiable polluters can be brought to court.

Hence, common-law protection of property rights does not seem to provide a reasonable or effective solution for the many millions of people who are both victims and perpetrators of pollution, as in the case of smog. In such cases, if the government is to protect its citizens and their property against invasion and harm from others, including polluters, it must sometimes regulate more directly.

Partly because of these problems, a shift occurred in the second half of the 20th century. Dissatisfaction with the way the courts handled pollution problems began to develop. Undoubtedly, this change also happened because people became more alarmed about pollution. The nation was getting wealthier, and people had food on their tables. Pollution bothered them more than it had bothered their parents and grandparents. From relying primarily on courts for protection against pollution, the nation moved to direct government regulation of polluting activities.

2. Direct regulation of polluting activities bypasses—but does not eliminate—the problem of missing information.

Protecting property rights in common-law courts was clearly an imperfect solution to the problems of pollution. As critics pointed out, the courts were slow and the outcome uncertain. Without better scientific data, the courts were not well equipped to deal with pollution that might cause illnesses such as cancer, which can be triggered at one point in time but not actually appear until many years later. Furthermore, the connection between chemical exposure and disease might be based on probability, not on a clear link. Knowledge and proof of cause and effect would be difficult to establish.

As mentioned in Chapter 1, around 1970 Congress began to pass a series of environmental laws that gave federal agencies sweeping powers to directly control activities that might have environmental consequences. The early 1970s were full of new laws such as the National Environmental Policy Act (1970), the Clean Air Act (1970), the Clean Water Act (1972), the Endangered Species Act (1973), the Toxic Substances Control Act (1976), and the Resource Conservation and Recovery Act (1976). In 1980, the Superfund law designed to clean up hazardous waste dumps was adopted.

Standards limiting pollution were set. Were they too tight or too lax? Would the best standards be different in different areas? Technologies are often specified in the regulations formed under such laws. Were they the right ones? Will they continue to be the right ones? Answering any of these questions requires information of the same kinds that courts would need to address the same problems. Yet the information is not necessarily produced. In fact, government agency officials may have little interest in gaining objective answers to the questions, preferring more and tighter regulations, at lower costs to themselves.

As a result of those laws, the mushrooming power of federal regulatory agencies, especially the Environmental Protection Agency (EPA), helped to reduce certain pollution problems. Environmental laws have had the strong support of most citizens, and environmental activists have become more influential as their claims have warned of large and imminent dangers ranging from acid rain to global warming. Their fundraising letters have been especially vivid, and the problems they have depicted proved to be valuable fundraising tools. Over time, more and more stringent regulations were adopted. At the same time, the costs imposed on taxpayers and on those forced to comply with regulations mushroomed.

Despite some early successes, many of the costly regulations, when closely examined by economists and other policy analysts, did not appear to yield large benefits. They were popular with a frightened public, but when better information appeared, often long after the "crisis" had ended, the benefits were often much smaller than had been expected. Frequently, dangers had been exaggerated and the solutions did not work as intended.

One example is the Comprehensive Environmental Response, Compensation, and Liability Act, known as Superfund. This 1980 law was designed to clean up hazardous waste sites and prevent future contamination from new ones. It allowed government officials to pursue their narrow goals without taking into account competing goals or having to provide the kind of cause-and-effect information required in the courts. Implementation of the law resulted in costly programs that produced little in the way of demonstrable benefits.

Such laws led to something of a backlash. Much of the period after 1970 has been characterized by hostile confrontations. On the one side, environmental activists press for tougher regulations; on the other side, the companies and individuals who most obviously bear the burden of those regulations resist. Of course, many of us pay for inefficient regulation without realizing that the cost of what we buy and what we do is greater because of those regulations—with little benefit in return.

3. Decisionmakers in government agencies often fail to see the big picture; good intentions can lead to bad results.

It is true that markets are imperfect and that governments can sometimes solve specific problems. But government programs often end up helping organized and active political and bureaucratic constituencies (that is, special interests) rather than solving a problem in the broad interest of the unorganized general public. There are several reasons for this outcome. One of the most persistent factors can be summarized as tunnel vision. This is the term that Supreme Court Justice Stephen Breyer applied to federal regulators, including the EPA.[2]

For Breyer, tunnel vision is the tendency of government employees to focus exclusively on the objectives of their agencies, or even the specific programs within their agencies, at the expense of all other concerns.

As noted in Chapter 2, all people have narrow goals. Narrow goals lead to tunnel vision if the legislators writing the law are ambitious about the ends but do not provide guidance about the means of implementing the law. This is frequently the case.

The benefits of tighter regulations should be balanced against the sacrifices of resources made, or production forgone, to comply with stricter regulations. If Congress insisted on such balance, elected officials would have to make difficult decisions, but the results would be positive: agency officials would have to consider the real costs that regulations impose.

In contrast to regulators, who sometimes have expansive authority under vague laws, market participants can regulate in only two ways. The first way is to show in court that actions harming them are violating their rights, and that the court should halt the harmful actions, require cleanup, or insist that damages be paid. The second way is to lease or purchase rights. Companies along the Tar-Pamlico River Basin in North Carolina took this approach. The basin was being polluted by discharge from the farming in the area. The companies formed an association and used their fees to compensate farmers. In return, the farmers created buffers along creeks and cleaned up hog waste by using it for fertilizer.[3]

A regulator can write very strict and costly standards under some environmental laws with only a plausible claim of potential risks or potential gains. No proof of supposed risks or supposed benefits is necessary, and no review of regulators' decisions is required, so long as the regulators cannot be proven to have acted in an arbitrary or capricious fashion.

Not surprisingly, tunnel vision sometimes leads to excessive regulation that may cause more harm than good. The 1980 Superfund law, for example, created a large fund of tax money for cleaning up abandoned waste sites. The fund came initially from a tax on chemical-producing industries, but the EPA was authorized to obtain compensation from *any* individual or company that had deposited any hazardous waste in the sites. To obtain this compensation, EPA officials have no responsibility to show wrongdoing or actual damage to others or even any real and present risk emanating from the sites.

Superfund was aimed at a genuine problem. Buried wastes sometimes leaked from their underground sites and caused harm. If no owner responsible for the wastes was found who could be sued and forced to clean up the site and compensate those harmed, Superfund was supposed to get the job done.

The Love Canal waste site in Niagara Falls in upstate New York spurred passage of the law. In the late 1970s, chemicals from the site had leaked into the yards and basements of nearby residents, creating unsightly messes and bringing unwanted smells. And much worse was feared. Two hastily written health studies suggested that diseases and birth defects might have resulted from the leaking chemicals. The fact that the studies were soon discredited was hardly noticed in the uproar.[4]

The Love Canal event galvanized Congress. Noting that the courts had not been able to bring swift, sure, low-cost relief to those who feared injury from chemicals leaking from underground storage sites, Congress passed the Superfund law. President Jimmy Carter signed it in the last weeks of his presidency.

But Superfund did not prove to be the swift, sure, low-cost solution that people wanted. It was supposed to cost at most a few billion dollars and to be paid for mainly by those whose pollution had caused serious harms or risks. But that was not the result.

In the first 13 years after Superfund was established, the program spent $20 billion, and its costs grew, along with delays in its cleanups of hazardous waste sites. Despite the expenditures, the program showed little gain in the way of human health benefits. In a 1996 study, *Calculating Risks*, researchers James T. Hamilton and W. Kip Viscusi reported a number of findings that challenge the received wisdom about the Superfund program.[5] Among them were the following:

- Most assessed Superfund risks do not pose a threat to human health now; they might do so in the future, but only if people violate commonsense precautions, actually inhabit contaminated sites, and disregard known risks.

- Even if exposure did occur, there is less than a 1 percent chance that the risks are as great as the EPA estimates, because the EPA makes extreme assumptions and then accumulates them.

- Cancer risk is the main concern at Superfund sites because cancer has a long latency period and some contaminants at the sites can cause cancer in high-dose exposures. Yet without any cleanup, only 10 of the 150 sites studied were estimated to have one or more expected cases, and at most of the sites, each cleanup is expected to avert only one-tenth of one case of cancer.

- In their study, replacing extreme EPA assumptions about the dangers with more reasonable ones brought the estimated median cost of preventing a single case of cancer to astronomical levels. At 87 of the 96 sites that had the necessary data available, the cost per *cancer case averted* (only some of which would mean a life saved) was more than $100 million.

- Given that other federal programs commonly consider a life saved to be worth between $7 million and $9 million,[6] this figure means that diverting expenditures from most Superfund sites to other sites or other risk-reduction missions could save many more lives—or save the same number of lives at far less cost.

The Superfund story is a clear case of tunnel vision. EPA site managers have little reason to worry about whether money spent by others to reduce environmental risks at Superfund sites caused other important social goals to receive less money. Agency officials have pushed cleanups beyond the point of efficiency. Hamilton and Viscusi estimate that 95 percent of Superfund expenditures are directed at the last 0.5 percent of the risk. As President Bill Clinton said in 1993, "The Superfund has been a disaster."[7] Most observers agree.

When agencies such as the EPA have the authority to make demands that they do not have to pay for and do not have to justify, efficient decisions are unlikely. Other agencies also face the same temptation— to put enormous demands on the private sector simply because tunnel vision tells them to do so and no budget restraints hold them back.

Some agencies take additional steps. Committed to their narrow goals, administrators will use their expertise, their control of information about programs, and often their monopoly position to push for a greater budget to pursue their mission. For example, the National Park

Service has often used what observers call the Washington Monument strategy.

The strategy goes like this: When the federal budget is being formulated, the National Park Service (like other bureaus and agencies) usually proposes a large increase, which then is trimmed by the Department of the Interior, the Office of Management and Budget (OMB), or the relevant congressional committee. The agency's budget, at least in recent times, has always been increased over the previous year, but the increase is always smaller than the park service would like. One response by the park service has been to announce that it may have to economize by shortening the hours that it can operate the Washington Monument—or another especially popular attraction.

The threat of long lines of citizens (i.e., voters) waiting to get in, outraged at not being able to enter, often persuades congressional committees or political appointees in the OMB to increase funding. In effect, park service leaders are saying, "Give us what we asked for, or we will cut back on our most popular services." The strategy tends to enlarge the park service budget.

Private firms rarely use that strategy or anything like it. Can you imagine Walmart coping with lower revenues by reducing its services, dropping its most popular product lines, and shortening the hours of its most popular stores—and then advertising that this will continue until customers give the chain more business?

The Washington Monument strategy would not work if the park service were depending on user fees for its revenues. The agency would not shorten the hours at the most popular visitor site—that would reduce revenue more than shortening hours at less popular sites. A private store facing budget problems would cut its *least* popular store hours, not the most popular ones. A firm competing for customers has to recognize that customers always have other options. If a firm serves its customers poorly, few will return and revenues will decline further.

The strategy used by the private sector is exactly opposite to that of a government agency such as the National Park Service. The agency knows that Congress must respond to complaints from voters and that taxpayers must pay any increased taxes that Congress might levy to increase the agency's budget.

Every agency (federal, state, and even local) is doing something similar to compete in the budget process. In the case of federal agencies, most are working with sympathetic congressional committees.

Although competition for budget funds is great, the upward pressure on the total budget also is strong.

So government regulation is imperfect. However, this does not imply that government should never act beyond the protection of property rights by common law. Instead, it means that when we turn to government to address environmental issues, we should expect to encounter problems as well as hoped-for solutions.

4. It is understandable that the individual voter may be uninformed about most policy matters and even about elected representatives.

By and large, voters do not monitor and correct the problems of government control. For various reasons, voters, who are often intelligent and well intentioned, remain ignorant about most issues and therefore are not successful in monitoring their elected representatives.

First, voters seldom decide policy issues directly. Instead, they vote for political candidates. And they frequently lack the detailed information needed to cast their ballots for candidates in a truly knowledgeable fashion.

In most elections, a single voter's choice is not decisive. Voters know this. Recognizing that the outcome will not depend on one vote, the individual voter has little incentive to spend time and gather information on issues and candidates to cast a more informed vote. This factor explains why most Americans of voting age cannot name their congressional representative even after the election, much less identify, understand, and compare the positions of candidates on environmental issues—or most other issues—most of the time.

To grasp in a more personal way why citizens are likely to make better-informed decisions as consumers than as voters, imagine that you are planning to buy a car next week and also to vote for one of two candidates for the U.S. Senate. You have narrowed your car choice to either a Ford Taurus or a Honda Accord. In the voting booth, you will choose between candidates Sam Smith and Amanda Jones. Both the car purchase and the Senate vote involve complex tradeoffs for you. The two cars come with many options, and you must choose among dozens of different combinations. The winning Senate candidate will represent you on hundreds of issues, but you can choose only one candidate.

Which of these complex decisions will command more of your scarce time for researching and thinking about the best choice? Because your car choice is entirely yours and you must pay the entire cost of what you choose, an uninformed car purchase could be very costly for you. But if you mistakenly vote for the wrong candidate out of ignorance, the probability is nearly zero that your vote will decide the election. Your individual vote almost never controls who will actually win. Cumulatively, your vote and those of all the other people in your state will decide the election, but your choice will not. You recognize that a mistake or poorly informed choice will have little effect on the actual outcome.

It would not be surprising, then, if you spent substantial time considering the car purchase and little time becoming informed about the candidates or the political issues. Automobile choices are not perfectly informed decisions, but the buyer is certain to benefit from giving careful consideration to the alternatives. As a result, automakers are probably guided by better-informed votes (dollar votes, that is) than the U.S. Senate, even though decisions made by the Senate are far more important for the nation as a whole than automobile choices are.

The fact that voters have little incentive to study issues and candidates carefully has enormous ramifications. First, voters will rely mainly on headlines in newspapers and brief, sound-bite-length television reports, paid political advertising, Internet postings, and other information they can pick up casually. Voters do not encourage media to spend time and space on the detailed and complicated information that would be necessary for them to make informed decisions. What sells well on television and in the news are the human-interest stories about villains and heroes, with dramatic images of shocking, high-risk situations.

That is what happened with the 2010 BP oil spill in the Gulf of Mexico. It started with a horrendous explosion that killed 11 workers. Then, oil gushed from the Deepwater Horizon drilling rig for nearly three months. But as Stephen Moore and Joel Griffith pointed out in an article in the *Daily Signal*, the results are invisible today. Five years after the spill, say Moore and Griffith, "the lasting ecological damage from the spill that was originally feared, has not happened. The dire predictions by the media and the major environmental groups proved wildly off base." They continue:

> Today, the Gulf region affected by the spill is enjoying a renaissance of energy production, booming tourism, and a healthy fishery industry. Scientific data and studies over the past five years show the Gulf environment is returning to its baseline condition. The remnants of the spill are hard to find.[8]

Voter ignorance (albeit rational) lays the foundation for laws that attack villains and have high-sounding goals. By the time the laws are implemented, the voter has turned to other matters, especially because the details are complex and not very interesting. The ignorance of the voter explains why Superfund and the Endangered Species Act were popular when they were passed, why most voters know little about the programs' problems, and why the voters have not success-fully demanded that their elected politicians eliminate the problems. This ignorance is duplicated in many environmental issues.

5. Government has no capital market, so it lacks the signals and the incentives associated with market decisions.

As noted earlier, decisions made by private firms are evaluated in the private sector's capital market. Because such a capital market is missing when it comes to government ownership and control, gov-ernment officials do not receive correct and persuasive signals about whether their management is sound.

In a private market, when investors view a management decision as a good one, they keep their stock or buy more, anticipating that the value of the firm will rise. If many investors begin to think this way, their decisions lead the stock's price to rise, making its shareholders instantly more wealthy. Similarly, poor decisions lead shareholders to sell the stock, and the price tends to fall. Managers respond to the capi-tal market signals about the fluctuating value of their stock by taking appropriate action. Those who do not are likely to be replaced.

Government managers do not get capital market signals. The lack of signals causes numerous difficulties. It poses a special problem because the federal government owns about one-third of the land mass of the United States but does not get any guidance from capital markets on whether it is managing those properties well. Government forests, grasslands, wildlife preserves, parks, and other resources (including government buildings) are seldom sold, so there is no price signal established through market trades as there is for resources in the private sector. Changes in government policy or events such as forest fires do change the value (to society) of a government-owned resource, but the changes are not reflected in any market price.

Furthermore, unlike investors in the private sector, few citizens have a strong financial incentive to learn what is happening with the management of government-owned resources. Because the assets are seldom bought or sold, no one benefits directly from knowing about management changes. So there is no feedback to decisionmakers from resource price changes of the kind that private owners and managers receive. In contrast, when an owner or manager in the private sector takes actions that affect the resource value—such as Ted Turner's offering expensive elk hunts on his ranches, as mentioned in Chapter 3—the market value of the resource changes (in Ted Turner's case, going up significantly).

Government agencies operate without such information and without the system of rewards and penalties that a market for capital assets passes along to private decisionmakers. Rewards and penalties give signals and incentives to properly plan for the firm's future, even at the expense of today's profits or dividend payouts.

6. Special-interest groups try to use the government's resources and regulatory authority to further their own narrow purposes.

By its very nature—because it has the power through taxing and spending and through regulation to coerce people to take actions, rather than relying on choice—government is called on to take from some to give help to others. A government decision often generates substantial personal benefits for a small number of constituents while imposing a small individual cost on a large number of other voters. The big benefits to the small number of recipients provides them with an incentive to lobby hard. Yet the rational ignorance of the voter (as previously discussed) means that most of those who pay the costs are unaware that they are doing so.

The federal government's program to supply below-cost water to farmers in the West illustrates this imbalance. Because the West has very limited rainfall, the federal Bureau of Reclamation has built dams to create reservoirs and irrigate fields. In Utah, the Bureau of Reclamation used the Central Utah Project's dams and canals to deliver irrigation water from a tributary of the Colorado River to Utah farmers. This transfer of water was highly subsidized by the federal treasury. The price to the farmers was only $8 per acre-foot (enough water to cover an acre one foot deep) even though the cost of the delivered water was about $400 per acre-foot. Estimates put the value of the water to farmers at about $30 per acre-foot.[9]

The below-cost water delivery served the landowners and farmers and the small communities where they live. The high costs of dam construction and operations (well above the amount the farmers paid) were passed on to taxpayers across the nation. Because each individual taxpayer paid only a fraction of the total cost, to this day most taxpayers have never heard of the project and have no idea of the costs they paid.

Environmental regulations also are frequently influenced by business firms and unions seeking protection from competitors in the marketplace. The Clean Air Act Amendments of 1977 are a classic illustration of how factional political interests can override the public interest.

The amendments required coal-burning electric power plants to install expensive "scrubbing" devices in their smokestacks to reduce sulfur dioxide emissions from the exhaust. However, in many cases, the emissions of sulfur dioxide could have been reduced merely by using cheap, low-sulfur coal instead.

Because coal companies and their unions in places such as West Virginia and Kentucky produce high-sulfur coal, they didn't want competition from low-sulfur coal, most of which comes from the West. So, working through their political representatives, they froze out the competition by insisting on scrubbers, which won't even work when used with low-sulfur coal (some utilities had to add sulfur to make them work!). Most citizens didn't realize that they would pay more for electric power because of these regulations or that their air would be a little dirtier than if the companies could use low-sulfur coal.

A number of factors combine to make special-interest groups far more powerful in a representative democracy than their numbers would indicate.[10] Members of an interest group—such as the owners of specific tracts of farmland irrigated with low-cost water—have a strong stake in the outcome of some political decisions. Thus, they have an incentive to hire lobbyists to help them in Congress and with regulatory agencies. They also have an incentive to inform themselves and their allies in local communities and to let legislators know how strongly they feel about an issue of special importance. Many of them will vote for or against candidates for election strictly on the basis of whether the candidates support their specific interests. Such interest groups are also in a position to provide campaign contributions to candidates who support their positions.

Environmental groups are special-interest groups, too, but they are different in that they have enough friends in the media that they can

59

often rally the public. Hence, they can have an important voting base, as illustrated by the current opposition to the Keystone XL Pipeline. This pipeline would be built to carry shale oil from Canada to the Gulf of Mexico. But a coalition of environmental groups, including the Natural Resources Defense Council and the Sierra Club, oppose the pipeline. They don't want more oil to be produced, especially from Canada's tar sands, which could provide an almost unending supply of oil.

This coalition was able to persuade President Barack Obama to delay (perhaps indefinitely) the construction of the pipeline, even though it had passed numerous environmental assessments and had the backing of important union groups.[11]

Because most voters know little about someone else's special-interest issue, examining the issue takes much more time and energy than it is worth in terms of possible personal gain from eliminating a subsidy or other special help. Of course, there are many such issues, but each would have to be considered separately, so most voters who are not special interests simply ignore them.

If you were a vote-seeking politician, what would you do? Clearly, regarding an issue on which a special interest is taking a position without organized opposition, little can be gained from supporting the interest of the majority, which is largely uninformed and therefore uninterested. Supporting the position of the well-organized group can generate vocal supporters, campaign workers, and importantly, campaign contributions—even though that position may be contrary to the interests of the public as a whole.

The ability of the voter to punish politicians for supporting costly special-interest legislation is further hindered by having many issues "bundled" together when the voter chooses between one candidate and another. Even if the voter knows and dislikes the politician's stand on one or a few issues, the fact that hundreds of future issues are bundled into each candidate will severely limit the voter's ability to take a stand at the ballot box for or against any particular issue.

Officials of government agencies—bureaucrats—also favor many special-interest programs. The bureaucrats who staff an agency usually want to see their department's goals furthered, whether the goals are to protect more wilderness, build more roads, or provide additional subsidized irrigation projects. Accomplishing these goals requires larger budgets and staffs. Not so incidentally, the programs provide the bureaucrats with expanded career opportunities while helping to satisfy their professional aspirations as well. Bureaus, therefore, are usually eager to expand their programs to deliver benefits to special-interest groups because the groups will work with politicians to expand the bureaus' budgets and programs.

7. Government policies that erode the protection of property rights reduce the ability and the incentive of owners to protect and conserve their resources.

Many environmental policies erode property rights. When they do, they often work against their stated goal of protecting the environment. The unintended results can sometimes be dramatic.

The Endangered Species Act (ESA), intended to save species thought to be in danger of extinction, is an example. Although few species have actually become extinct since the act was passed, only 28 of the 2,105 species listed as endangered or threatened have been removed from the list because their species or populations have recovered.[12] Thus, the act is not a success story by any measure. The far-reaching powers vested in federal agents to control the landowners' use of their properties have sometimes worked to protect endangered species, but often they have had the opposite effect.

How can that be?

A landowner who provides a good habitat for a listed species, even by accident, is likely to lose the right to use the land as he or she wishes. Michael Bean, an environmental attorney who is sometimes credited with authorship of the Endangered Species Act, made this point some years ago. Speaking to a group that included U.S. Fish and Wildlife Service (FWS) officials, he said that there is "increasing evidence that at least some private landowners are actively managing their lands so as to avoid potential endangered species problems." He emphasized that these actions are "not the result of malice toward the environment" but "fairly rational decisions, motivated by a desire to avoid potentially significant economic restraints." He called them a "predictable response to the familiar perverse incentives that sometimes accompany regulatory programs, not just the endangered species program but others."[13]

The case of Benjamin Cone Jr. is a cautionary tale.[14] Cone inherited 7,200 acres of land in Pender County, North Carolina. He managed the land primarily for wildlife, planting chuffa and rye for wild turkey, for example. The wild turkey has made a comeback in Pender County

61

partly because of his efforts. Cone also frequently conducted controlled burns of the property to improve the habitat for quail and deer.

Red-cockaded woodpeckers are listed as an endangered species. They nest in the cavities of old trees and are attracted to places that have both old trees and a clear understory. By clearing the understory to protect quail and deer and by selectively cutting small amounts of timber, Cone may have helped attract the woodpecker. Cone knew that he had at least a couple of red-cockaded woodpeckers on the property.

When Cone intended to sell some timber from his land, the presence of the birds was formally recorded by the U.S. Fish and Wildlife Service. The agency warned Cone not to cut trees or take any other actions that might disturb the birds. They did not, however, tell Cone where the nests were. Cone hired a wildlife biologist, who estimated that there were 29 birds in 12 colonies. According to the FWS guidelines then in effect for the red-cockaded woodpecker, a circle with a half-mile radius had to be drawn around each colony, within which no timber could be harvested. If Cone harvested the timber, he would be subject to a severe fine, possible imprisonment, or both under the ESA.

Biologists estimated that the presence of the birds and the FWS rules put 1,560 acres of Cone's land under the restrictions of the U.S. Fish and Wildlife Service.

In response, Cone changed his management techniques—in the opposite way the law had intended. He began to clear-cut 300 to 500 acres every year on the rest of his land. He told an investigator, "I cannot afford to let those woodpeckers take over the rest of the property. I'm going to start massive clear-cutting. I'm going to a 40-year rotation instead of a 75- to 80-year rotation."[15]

By harvesting younger trees, Cone could keep the woodpecker from making new nests in old tree cavities. He also took steps to challenge the FWS in court, asking to be compensated for his losses. In response, the agency avoided that court challenge by negotiating a settlement that gave Cone more freedom to use his land.

Cone's experience teaches a lesson to all landowners who learn about his situation. They may be in for similar treatment unless they do something about it. Indeed, after Cone informed the owner of neighboring land about possible liabilities in connection with the red-cockaded woodpecker, he noticed that the owner clear-cut the property.[16]

Overall, what has been the result of the ESA for the red-cockaded woodpecker?

As Bean has said, "The red-cockaded woodpecker is closer to extinction today than it was a quarter century ago when protection began." Bean recommended that the rules be changed to help landowners avoid large reductions in the value of their land from the application of the ESA. He has worked with the federal government to let landowners create limited "safe harbors": landowners will not be punished for having endangered species on their property, but in return they must work with the government to enhance the habitat.[17]

As mentioned in Chapter 3, the Delta Waterfowl Foundation has a low-cost program that pays small amounts to farmers who take extra effort to protect nesting areas for ducks. What would happen if the ducks became listed species? It seems safe to say that farmers who now cooperate to help the ducks would be wary of enticing listed species onto their lands. Even with the payments, would an owner risk the loss of control on much of the farm?

By using the Endangered Species Act to justify land-use controls that seriously erode the property rights of landowners, the U.S. Fish and Wildlife Service has ignored the important positive role that private landowners and institutions have historically played in protecting rare fauna and flora.

8. When the government promotes the goals of some citizens at the expense of others, resources are diverted from production into political action.

As we have seen, the government sometimes forces the transfer of resources from some groups to others without compensating the losers. Interest groups have learned that they can benefit by influencing these transfers. That process gradually causes a shift in the economy. Hiring lobbyists to influence laws and tax experts to find loopholes becomes more lucrative than hiring innovative scientists, engineers, and production personnel. The output-expanding, positive-sum activities of market discovery, innovation, and production are increasingly replaced by resource-consuming, negative-sum battles to gain political transfers or avoid paying for transfers to others.

As transfers that depend on political clout increase, more people redirect their energies to clout, taking away more time, energy, and other resources from productive activities. Competition tends to shift producers' energy and attention from innovations in production and trade to competing for political favors.

When political redistribution of society's goods and services (the pie) grows, fighting over shares reduces efforts devoted to increasing the size of the pie, thus making it smaller. As the political stakes have grown, resources have been increasingly diverted toward lobbying activities. The number of registered lobbyists in 2013 was approximately 12,000. James Thurber, a university professor who helped design the Obama administration's lobbying rules, suggested that the actual number of people lobbying Congress or the executive branch is around 100,000.[18]

According to the Competitive Enterprise Institute, Americans spent $1.83 trillion complying with federal regulations in 2013.[19] Environmental regulation was an important part of those regulations—by one measure, it represents about 30 percent of "economically significant" regulations.[20]

The stated purpose of regulation is seldom to transfer wealth from one group to another. But regulation almost always does just that. We saw earlier in this chapter that stiff environmental regulations for new power plants (the "scrubbers" requirement) helped regions with high-sulfur coal keep out competition from low-sulfur coal mines, primarily in the West. By forcing firms to bear greater regulatory burdens, environmental regulation often helps the competitors of those firms.

Economist Bruce Yandle has developed a theory that offers one reason that the transfer of wealth occurs. It's called "bootleggers and Baptists," a term based on his scenario that explains why "blue laws" continued in the southern United States into the middle of the 20th century. A blue law might prevent the sale of liquor on Sundays. Baptists (or any other group opposed to alcohol) would make a public case for avoiding alcohol on the Lord's Day. The businesses that produced or sold alcohol illegally (bootleggers) quietly lobbied the legislature for the same laws—because those laws gave them a good market.

An illustration of Yandle's theory is the lobby for government policies that favor the fuel ethanol, which is used in gasoline and is made from corn. Environmentalists espoused ethanol as a "natural" fuel and an alternative to fossil fuels (even though it has some negative environmental impacts that may not have been recognized initially). They are the "Baptists" in the scenario. The bootleggers are the farmers and farm organizations that obtained higher profits because of government subsidies and special treatment.[21]

Efficient use of resources is not the only victim of increased transfer activities. The legitimacy of government may suffer when officials increasingly tax some citizens to transfer payments to others or when agencies transfer the use of government lands from groups that have less political influence to other more politically potent groups.

In the San Joaquin Valley of California, for example, a small fish, the delta smelt, has pitted environmentalists against farmers.[22] In 2006, the Natural Resources Defense Council went to court, arguing that the delta smelt must be saved under the Endangered Species Act. The organization was successful, and the federal government is preventing billions of gallons of water from going to farmers, to protect the delta smelt. After years of drought, farmers who can't grow crops are angry that so much water has been cut off for the sake of a three-inch-long fish.

Those who lose income or lose access to resources without compensation are often upset. For them and for others who do not benefit from the transfers, these transfer programs make the public-interest rhetoric of government action seem hollow. Political battles over economic benefits for one group at the expense of others create ill will, which in itself can harm the public welfare.

9. The government's environmental monitoring services can help provide information about the environment that property rights and markets might not produce.

Despite heavy intervention in the realm of environmental regulation, the federal government has done little in an area where its contribution could be critical: the collection and preservation of data. Whatever system is responsible for controlling pollution—whether property rights and common-law or government regulation—good data are needed. It is necessary to know which pollutants are causing harm, where the pollutants are coming from, and whether pollution levels are improving or getting worse.

Government-maintained monitoring systems, such as a network of sensing equipment that records the level of pollution at many locations, can provide geographically detailed information. Such a network can help explain whether changes have resulted from the addition or subtraction of effluents from specific sources, from new industry mixes, or from other origins. Such information can help epidemiological researchers trying to learn when pollutant exposure at various levels is correlated with harms to people and property and when it is not. Researchers may observe what they think are the effects of pollutants, but without accurate data they have trouble connecting pollution levels with harmful effects.

The federal government has received much criticism for not doing enough to learn about actual pollution levels. Debra Knopman, an environmental scientist and former U.S. Interior Department official in the Clinton administration, put it this way:

> Imagine controlling the heat and air conditioning in a 50-room mansion with one cheap thermostat, or pulling smoke detectors off the mansion walls to save on buying batteries. Compared to the cost of wasted electricity or damage from fire, such penny pinching on monitoring temperature and smoke in the mansion is simply absurd. Yet, this is precisely what we do when we regulate the environment while so poorly monitoring our progress or keeping tabs on how conditions in the air, water, and land are changing over time.[23]

Among the examples that Knopman notes is a sparse and underfunded network for monitoring water quality; it can barely tell us anything about progress under the Clean Water Act. Nor does a national monitoring network exist to measure small particulate matter in the air, even though the Environmental Protection Agency has set stringent standards for those particulates.

Regulators need these data to make rational decisions. The common-law protection of property rights against harmful pollution would also be well served with such data. Individuals who fear they have been harmed by pollution and those accused of producing harmful pollution who contend they are innocent both would value this information.

Expanding government data gathering of ambient pollution levels would seem to be quite useful, especially because regulators are in fact making decisions that could benefit greatly from information not currently being gathered. But because government decisionmakers do not personally face the costly consequences of poorly informed decisions, they have little incentive to ensure that accurate information is obtained.

10. Competition is important in government processes, just as it is in markets.

We have seen that state parks depend partly on consumer support. That situation means that they have to compete for the consumer's dollar, which can easily be spent elsewhere. Once they became more reliant on user fees, state parks found ways to enhance the services they provided. The U.S. Forest Service has experienced something similar.

For years, environmentalists complained that the Forest Service was excessively influenced by the timber industry. Instead of using its budget to provide trails and campsites to serve the growing number of people who hike and camp in the national forests, the Forest Service emphasized logging. In the 1990s, Randal O'Toole, an environmentalist and forest economist, argued that the solution to this problem was competition. The Forest Service should start charging fees to people who hike in the woods. Those fees, he said, would give Forest Service officials an incentive to do more for hikers and backpackers, including perhaps avoiding some of the clear-cuts they objected to.

Some of this change has come about. As part of a federal program that raised or introduced fees in four government agencies, the Forest Service began earning funds by better serving consumers. The law required that the funds be used in the Forest Service units in which they were received; some funds are going into rehabilitating trails and providing camper services. In essence, recreation (which consumers are paying for) is competing with timber, changing the Forest Service's budgetary allocations.

Competition, whether it affects government agencies or private firms, is a disciplinary force. When customers have choice, poor goods and services cause providers to lose business to rivals that offer a better deal. Thus, competition protects consumers against high prices, shoddy merchandise, or poor service. We generally recognize this point in the private sector.

Competition in the public sector can be equally important. The current incentive structures facing most government agencies and enterprises do not reward efficient operation. The directors and managers of public-sector enterprises seldom gain by working to reduce cost and improve performance. In fact, if an agency adopts cost-cutting measures and thus fails to spend all of one year's budget allocation, it has a weaker case for keeping or enlarging its budget for the next year. Agencies typically spend all of each year's appropriation, even if it means spending a lot on low-priority items late in the budget period. In effect, inefficiency is rewarded with a larger budget.

In the private sector, the profit rate is a measure of how much value was added relative to the purchase cost of resources used. Profit provides a clear index of performance. In a competitive market, when property rights are protected, profit indicates that resources were purchased at a price lower than the resulting product was worth to buyers. In contrast, loss indicates that the product was worth less than the resources taken from the rest of the economy to produce it. In the private sector, low rates of profit or bankruptcy eventually weed out inefficiency.

But there is no indicator of performance such as profit in the public sector, so managers of government firms can often continue despite economic inefficiency. There is no mechanism in the public sector that parallels private-sector bankruptcy or withdrawal due to low profits, thus ending wasteful programs. In fact, poor performance and failure to achieve objectives are often used as arguments for increased funding in the public sector. As the Washington Monument strategy illustrates, National Park Service administrators use poor service to visitors (or the threat of poor service) to argue for increased funding. Every agency uses some form of this method simply to protect and enhance its budget.

With such incentives, it is vital that government units face competitors. The competition will improve performance, reduce costs, and stimulate innovative behavior. As a result, waste of resources can be reduced, and citizens will get more for their money.

One way that competition is introduced in governments is to force agencies to seek part or all of their budgets from user fees, as the Forest Service does to a limited extent. Another way is to decentralize decisionmaking so that states and municipalities (rather than the national government) can decide how to spend funds. Decentralization works because citizens can vote with their feet.

Just as people differ on how much they want to spend for housing or automobiles, so too will they have different views on how much to sacrifice for environmental benefits. Some will prefer to live in towns or cities with a high level of environmental services and will be willing to pay higher taxes for them. Others will prefer lower taxes and lower regulatory burdens, along with fewer environmental services. A decentralized system can accommodate these divergent views.

Competition among local governments will also promote government innovation and efficiency. When citizens can easily vote with their feet, governments will have a greater incentive to provide services economically. If a government regulates in a costly way (when lower-cost methods are available) or levies high taxes without providing a parallel value to voters, both individuals and businesses will be repelled and will go elsewhere. Similarly, when people bear burdens for items that provide them little or no value, many will choose the exit option.

Thus, like business firms in the marketplace, local governments that fail to serve their citizens will lose customers—that is, population—and tax revenues.

Competition among decentralized governments serves the interests of the citizen. If competition is going to work, however, the policies of the federal government must not stifle it. When a central government subsidizes, mandates, and regulates the bundle of services provided by local governments, it undermines the competitive process among them.

Conclusion

We cannot expect government managers, if given the authority, simply to fix the shortcomings of the court-enforced property rights and market trading system. They may be able to make some contribution, but they will experience incentives and information problems that lead to unintended consequences.

When the government uses its power to increasingly tax, spend, and regulate, more groups try to turn that power to their own narrow advantage. For most citizens, the results are not likely to improve their situations and may instead worsen them.

Yet the fact remains that for a few kinds of environmental problems, such as smog in the Los Angeles basin, enforcing property rights does not appear to be a feasible way to protect human health and other values, at least for now. When private enforcement is not available, coercive mechanisms may be the only alternative, for better or for worse. What criteria can help us to choose wisely between private and political management of the environment? Chapter 5 examines that question.

5. Choosing: Economics and Environmental Policy Choices

Two different economic systems can be used to address environmental problems. One system emphasizes politics and government organization. The other system relies primarily on private property rights and market relationships. Each depends on different signals to communicate what people want. And each provides different incentives for individuals to heed those signals.

This chapter further explores both the market mechanism and direct government controls, with the goal of evaluating the role of each. It begins with the market approach (property rights protection and exchange), which was used initially in the United States to deal with most environmental problems. More recently, the market approach has been supplemented with—and often supplanted by—the political approach.

Addressing environmental problems through markets occurs through the exchange of property rights (buying and selling property) and reliance on common-law courts and law officers. The government provides the courts and law officers, so the market approach also involves some political decisions because people vote for the judges or for the politicians who appoint the judges.

However, with the market approach, the government is in the background. The actual environmental results are largely determined by a combination of evolving technologies and evolving preferences of those who participate in the market and bear the consequences, both positive and negative, of their actions. This arrangement can limit harmful pollution, encourage conservation, and thus protect and enhance the environment.

In the past, the private rights and market option served as environmental policy in the United States and parts of Europe. Although these nations experienced periods of pollution and some harm to the environment, the market allowed for improvements in health, longevity, and other quality-of-life characteristics over several centuries. It also led to the control of severe pollution, the restoration of harmed areas, and the protection of large areas of land.

Efforts to preserve, enhance, and restore the environment increased as people in these countries became more affluent. Beginning with organizations like the National Audubon Society in the 19th century, the private protection of nature expanded throughout the 20th century (and continues today). Real estate developers paid more attention to preserving green space—and sometimes even habitat for wildlife—within residential areas, both in metropolitan areas and in resorts where people vacationed. Private organizations such as the Nature Conservancy and the Delta Waterfowl Foundation developed ways to protect endangered species and to preserve their habitat. Private companies such as Prairie Restorations (a company that preserves native vegetation and restores native environments) responded to the desire of many people to live near more natural or indigenous plants and flowers. Other innovators developed ecotourism—travel that showcases and protects natural environmental treasures such as the rain forests and wildlife. Adaptation and innovations are constant processes in a market system, and the growing interest in nature has spurred new ways to support environmental protection.

The market approach did not achieve all the environmental goals that many people wanted, however, especially when it came to air or water pollution. The market approach relies on innovation, changes in cultural habits, and common law. In the past, the changes were not fast enough for some, and the courts did not always protect individual rights against polluters or others who would misuse resources. Specifically, the courts could not fully address certain kinds of air pollution in big cities, where a great many polluters simultaneously harmed a great many citizens. Although air pollution declined, serious pollution problems remained. As a result of these shortcomings and for other reasons as well, direct government control has increasingly augmented or replaced the market (property rights) system over the past three decades.

Direct government control of the environment operates through two means. One is by regulating individual behavior through government agencies such as the Environmental Protection Agency (EPA). The other is through government ownership and management of resources, primarily land and water, by such entities as the U.S. Department of the Interior and state park agencies. The two means—government regulation and government ownership—may be combined, for example, when residents of a city are taxed to pay for a government-owned sewage treatment plant. In that case, the government requires citizens to contribute to the cleanup of waste, and it also owns the mechanism used for that cleanup. Either way, the approach is generally called "command-and-control environmental policy."

Since 1970, command-and-control has become the most frequently chosen type of environmental policy. In place of individual decisions and market trading, elected officials and agency appointees decide on the goals and often specify the means used to seek environmental protection.

Each system has strengths and weaknesses, and each modern industrial nation relies on a mix of the two approaches. As policymakers consider the future, they may have opportunities to change the mix. How much should we rely on property rights and market trading for environmental policy, and when should we turn to command-and-control instead? And can they work together? To answer these questions, consider these 10 key points:

1. Producing and protecting environmental quality are similar to producing and protecting any other good or service; individuals must receive accurate signals about what others want from them, and each must have incentives to heed those signals.

Many people think of the environment as different, yet producing and protecting environmental quality is similar to producing and protecting any other good or service. A forest, for example, can supply many goods and services. Those goods can include logs for houses and pulp for paper. A forest can also include streams for recreation and habitat for wildlife. Some of those goods and services—logs and, frequently, access to the forest for hunting, fishing, and camping—are traditionally sold on the market. Others, such as habitat for wild animals, are less often bought and sold through commerce.

Whichever economic system we choose to address environmental decisions must coordinate the desires of people who control the resources—whether government managers or private owners—with the desires of those who want to use them. Getting people to do what is needed for the good of others and getting them to refrain from wasteful or harmful acts require the right informational signals and the right incentives. And in the case of direct government control—where neither market prices nor the profit motive strongly guides users and producers to protect and preserve environmental quality—a means to force compliance with political decisions must be implemented.

2. Economic institutions matter. Economies around the world grow more rapidly and produce more goods and services per person, including environmental quality, when the role of government is smaller.

The productivity and wealth of nations depend as much on their *institutions*—the law, customs, incentives, and rules in place—as on their natural resources. Many countries, such as Argentina and Cuba, have rich natural resources but have had difficulty turning those resources into prosperity. Whether property rights can be traded, and under what conditions, can be more important than land or other physical aspects of a nation in determining how well people live.

Some people question this claim. They tend to think that if a country has good arable soil, it will produce adequate crops, or if it has plenty of rivers and ports, it will engage in trade. Through these natural resources, wealth will develop naturally.

One very simple comparison will challenge that assumption. Both Ethiopia and the Netherlands have arable cropland—about 15 million hectares in Ethiopia and about 1 million hectares in the Netherlands.[1] Ethiopia, a country with an authoritarian—although no longer officially socialist—government, produced about $452 worth of crops per hectare in 2012. The Netherlands, with a more limited government, produced $3,741 worth of crops per hectare that year.

Although a number of factors influence the productivity of Ethiopia and the Netherlands, the message that institutions can matter a great deal in producing agricultural output—more than available land in some situations—comes across loud and clear.

One comparison of this sort can be a good illustration, but it proves nothing. Even dozens of comparisons, randomly chosen, might not be persuasive, because most nations are not as easily classified into "market-oriented" or "nonmarket" nations as are the Netherlands and Ethiopia.

To compare the results of the market approach with the results of command-and-control more systematically, we need to know the extent to which markets and government control are used in each nation and then to compare characteristics of those nations. Economists James Gwartney of Florida State University, Robert Lawson of Southern Methodist University, and Joshua Hall of West Verginia University, using their own research together with that of many others, have constructed an index that enables us to do that.[2]

The Economic Freedom of the World index measures the degree to which private decisions and voluntary exchange drive the economy of each nation. The higher scores on this index reflect a greater role for private owners, a greater freedom to own property, a greater ability to receive protection for it under the law, and a greater ability to freely trade with others. The lower scores represent a greater degree of government decisionmaking in the economy.

The scores calculated by Gwartney and Lawson for 152 countries for 2012 range from 3.89 for Venezuela to 8.98 for Hong Kong. Countries having the smallest role for private property rights and voluntary exchange—the least amount of economic freedom by this measure—have the lowest scores. (Note that the economic system and economic freedom, rather than the political system and political freedom, are measured in this rating system.)

One way to use this index is to see how economic freedom or government control correlates with attributes such as a nation's worker productivity or agricultural output. When the 152 nations are ranked by economic freedom (with the top 20 percent or top quintile in one group, the next 20 percent in another group, and so on, down to the lowest 20 percent or lowest quintile), we find many favorable factors associated with more market influence and less government control over the economy.

For example, in 2001 Gwartney and Lawson published a chart that correlated World Bank figures on cereal production with relative degrees of economic freedom. Figure 5.1 shows that more economic freedom is associated with greater productivity of cereal grains per acre of good farmland. In other words, when government decisions play a larger role in a nation's economy relative to market decisions, a key indicator of agricultural productivity—cereal grain production per acre—goes down. (While the authors have not updated this particular example, the point remains valid.)

The implications of the economic freedom index are extremely broad. They go far beyond cereal grain production. As Gwartney and Lawson state, "Countries with more economic freedom have substantially higher per-capita incomes." In addition, they "grow more rapidly."[3] Figures 5.2 and 5.3 provide evidence that economies that are oriented toward free markets perform better economically than those with more government decisionmaking.

75

Figure 5.1
Cereal Crop Yields Rise with Economic Freedom

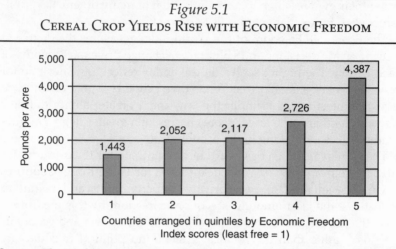

Countries arranged in quintiles by Economic Freedom
Index scores (least free = 1)

Source: The World Bank Development Data Group, "2001 World Development Indicators."

Figure 5.2
Countries with More Economic Freedom Have Substantially Higher Per Capita Incomes

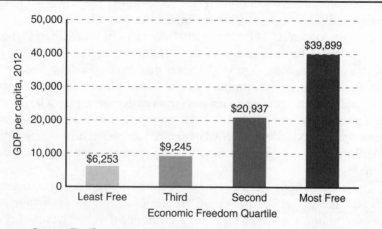

Economic Freedom Quartile

Source: James D. Gwartney, Robert Lawson, and Joshua Hall, *Economic Freedom of the World Annual Report 2014* (Vancouver, BC: Fraser Institute, 2014), Exhibit 1.6, p. 21, http://www.freetheworld.com/2014/EFW2014-POST.pdf.

Note: Countries arranged in quartiles by Economic Freedom Index scores. Dollars are per capita income in 2011 U.S. dollars.

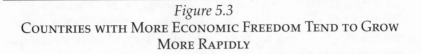

Figure 5.3
COUNTRIES WITH MORE ECONOMIC FREEDOM TEND TO GROW
MORE RAPIDLY

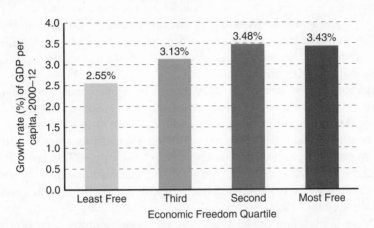

SOURCE: James D. Gwartney, Robert Lawson, and Joshua Hall, *Economic Freedom of the World Annual Report 2014* (Vancouver, BC: Fraser Institute, 2014), Exhibit 1.7, p. 21, http://www.freetheworld.com/2014/EFW2014-POST.pdf.

NOTE: Countries arranged in quartiles by Economic Freedom Index scores. The growth data were adjusted to control for the initial level of income.

For citizens in nations with higher rates of income growth and higher income levels, even the poor can live well. For example, in the United States, the income that is officially considered poverty level is more than twice the median income per person in the world.[4] In the United States, nearly three-quarters of households with poverty-level income have a car or truck. Eighty percent have air conditioning, nearly two-thirds have a cable or satellite TV, and more than half have a video game system such as Xbox. These statistics come from a report by Robert Rector and Rachel Sheffield, based on U.S. Census data.[5] The main point is this: economic institutions affect what people do and how they live even more than natural resources do.

3. Political and bureaucratic institutions tend to reduce efficiency and increase waste; the resulting decisions are less environmentally sound.

A dramatic illustration of what happens when ownership of land and its products is taken from private owners and given to the government followed the Russian Revolution of 1917. During the 1920s and 1930s, agricultural land was collectivized—seized by the government. Instead of working their own lands, farmers were assigned to work on giant collectives.

Before the revolution, Russia was known as the breadbasket of the world, producing more grain than any other country. But after the new policy was instituted, grain production fell sharply. Between 6 million and 11 million Russians starved to death, despite donations of food from nations with market economies.[6]

Because of widespread starvation, the government eventually allowed farm families each to have small private plots of less than one acre that they could farm either for their own use or to sell the produce in local markets. From these tiny privately owned plots, which added up to only 3 percent of the country's crop land, came 27 percent of the nation's food.[7] The Russian farmers, like other resource owners, were more willing to work hard and make productive use of a resource over which they had clear control.

After 1989, when the socialist nations began opening their borders and permitting more movement of goods and people, the income differences were obvious. So too were the differences in the environment.

Newspapers and magazines began reporting shocking examples, from drinking water seriously contaminated with arsenic in Hungary to pollution of irrigation water by heavy metals in Bulgaria. A giant pollution zone stretches out into the formerly pristine Lake Baikal, and the Aral Sea has drastically diminished in size because of irrigation for cotton plantations.[8]

Mark Hertsgaard, a writer who journeyed in China in the 1990s, describes visiting a paper mill in Chongqing. He saw a "vast torrent of white, easily 30 yards wide, splashing down the hillside from the rear of the factory like a waterfall of boiling milk." Suddenly, a gas explosion sent him running. This was a factory that was supposed to have been closed down because of excessive pollution—but wasn't. Hertsgaard also reported on the oppressive smoke in China's major cities.[9]

The socialist countries wasted resources, including energy. Studies by Mikhail Bernstam of the Hoover Institution found that market-based economies in western Europe used far less energy per $1,000 worth of output than the socialist nations of eastern Europe in 1986. Similarly, the European socialist economies used far more steel per unit of output than the European market economies did.[10]

The bottom line is this: control of resources by politics and bureaucracies does not bring the same pressures and personal incentives to innovate, to conserve resources, and to prevent damage downwind and downstream that private ownership and market decisions do. Political and bureaucratic decisions tend to be less efficient, more wasteful, and thus less environmentally friendly.

4. Protection of private property rights is associated with healthier environmental conditions and longer lives.

Recent studies show that in countries where property rights are better protected, people are healthier and live longer because of better environmental conditions. For example, using the economic freedom index discussed earlier, economist Seth Norton found that in countries where property rights are protected, 93 percent of the population has access to safe drinking water, whereas in nations with weak property rights, only 60 percent of the population has that access. Similarly, in nations with stronger property rights, 93 percent of the population has access to sewage treatment, whereas only 48 percent do in countries with weak rights.[11] Life expectancy is 70 years in nations with strong protection of property rights, whereas it is only 50 years in nations with weak or nonexistent protection.

One reason for these differences is that economically free nations are generally wealthier, as indicated previously. Certainly, wealthier nations have the wherewithal to take action that protects health, safety, and environmental conditions. Once people satisfy their basic needs, they begin to improve their environmental conditions.

There is more to it than that, however. Norton also conducted a study looking solely at the poor nations (countries with per capita income less than $5,000 in 1985).[12] He found that even among poor nations, 95 percent of the people live to age 40 in countries that offer relatively stronger property rights protection, whereas in the nations with weaker rights protection, only 74 percent of the people live to age 40. In rich nations or poor, property rights make an important difference.

When it comes to basic environmental protection—such as providing access to clean water—a system that protects private property rights is superior to one that relies on direct government control. Ownership of land or other assets gives the owner a legal right to use the courts to protect those assets. The owner has both the right and the incentive to protect and to find the most highly valued uses for the owned assets.

79

5. Replacing property rights (protected by the common law) with politically determined protection levels can result in lower environmental standards.

As noted earlier, the courts in a market economy are sometimes unable to protect rights against pollution. For example, if individuals cannot persuasively show the court that they are being harmed by pollution, the court will not stop polluters or make them pay damages. Failure by the courts to protect property rights can prevent the proper flow of incentives in the same way that failure to enforce laws and regulations of any sort can defeat the intent of the law.

That fact does not mean that government control will perform better.[13] One of the events that launched the modern environmental movement was the report in 1969 that the Cuyahoga River, which flows through the city of Cleveland and empties into Lake Erie, was so polluted that it burned. Of course, the water didn't literally burn, but there was oil on the water and lots of debris; a spark, probably from a train, ignited it. Public outrage at the thought that a river could go up in flames galvanized action and helped bring about tougher laws.[14]

It turns out that the Cuyahoga fire, which is still famous in some circles, occurred because efforts to obtain relief from river pollution through the courts had been replaced by command-and-control. A state pollution control board was in charge of issuing permits to emit pollutants into the water. The board had decided that a key stretch of the Cuyahoga was just an industrial river, so the companies along its banks did not have to clean up their effluent to any significant degree.

In fact, in 1965, Bar Realty Corporation, a real estate company, had tried to clean up a Cuyahoga tributary, but the Ohio Supreme Court concluded that the state pollution control board, not the courts deciding common-law claims, had the authority—and this board did not require cleanup.

Despite their imperfections, property rights, common law, and market relationships have some real advantages. Although judges and juries are not experts, in court they must listen to experts on both sides—each bound by rules of evidence and cross-examined by the other—before rendering a decision. That is far from the case when the same individuals (judges and jurors) enter an election booth to vote or when they vote as elected representatives.

As indicated in Chapter 4, voters are unlikely to be as informed as they would be had they been present through a trial of the facts, with its burden of proof, rules of evidence, and rights of cross-examination. With the exception of situations in which there are large numbers of polluters and victims, there is no obvious reason to believe that courts are less informed as they decide an issue than voters or even congressional representatives are on the same issue.

Evidence from Canada—where, as in the United States, statutory law and government control have been replacing decisions by private owners—suggests that the common-law protections are stronger. Researcher and writer Elizabeth Brubaker reviewed dozens of legal decisions and statutes. As political control supplanted the common-law approach to pollution, the protection of victims was weakened. In her book *Property Rights in Defence of Nature*, Brubaker writes:

> Governments have shown that they are not up to the task of preventing resource degradation or pollution; indeed they have often actively encouraged it. ... It is long past time for resources to be shifted away from governments and back to the individuals and communities that have strong interests in their preservation. Such a shift can best be accomplished by strengthening property rights and by assigning property rights to resources now being squandered by governments.[15]

Brubaker's book shows that property rights have been the better protector of environmental values over time. Individuals with property rights who face off against those who might harm them—governments included—will gain by finding ways to use rights effectively to protect themselves and their resources.

6. Environmental policies should be fair and cost-effective.

When government decisions are perceived as unfair, serious social conflicts can result. In the late 1980s and early 1990s, a property rights movement erupted around the country that was composed of voluntary organizations formed by people who believed that environmental regulations were violating their property rights.

Landowners—many of them small property owners who could not build on wetlands or who could not log their land because it contained endangered species—built this movement. In their view, individual property owners were being forced to bear the full cost of setting aside their land to produce habitat for wildlife and amenity values that benefited the public generally. They believed that the cost of such

81

production should be purchased from willing sellers and thus be more fairly distributed among the public. The owners of the land, they reasoned, should pay their share—and no more—of the cost of a habitat or environmental preservation.

This property rights backlash, in turn, caused a reaction among environmental groups, which viewed the property rights movement as an attack on environmental goals.

The mechanisms for protecting the environment should also be cost-effective for a very important reason. The willingness of citizens to pay for higher environmental quality depends on the cost. Inefficient policies cost more per unit of result, and they do not sell as well to voters as more efficient policies do. That may be one reason Congress rejected President Obama's 2010 proposal to tax carbon. (The EPA moved ahead with its own carbon-control policy in 2015.) When the costs are concentrated on a few people, more organized opposition to the conservation project is likely. Policies that achieve desired results without demanding huge economic or other sacrifices are an easier sell for those who care strongly about protecting the environment.

7. The change in pollution control policy from a property rights framework to administrative regulation is causing many Americans to die prematurely. ·

That is a strong statement, but it is supported by strong evidence. Many current environmental programs, including those designed to further health and safety, are accomplishing little at high cost. Tammy Tengs and colleagues at the Harvard School of Public Health studied 587 regulations and other federal government programs designed to save lives.[16] They found dramatic differences in the costs of saving lives and preventing illness and injury.

Their comparisons were based on the cost of extending one person's life by one year. Figure 5.4 shows some of the results when they compiled their numbers by agency. They estimate that the cost to save one additional year of life (or life-year) was $23,000 for Federal Aviation Administration regulations and $88,000 for Occupational Safety and Health Administration rules to reduce fatal accidents. Neither agency is especially noted for efficiency, but compare the median cost of their regulations to those of the Environmental Protection Agency. The EPA's regulations impose an estimated cost of $7,600,000 for each additional life-year extended.

Figure 5.4
MEDIAN COST/LIFE-YEARS EXTENDED

Federal Aviation Administration	$23,000
Consumer Product Safety Commission	$68,000
National Highway Transportation Safety Administration	$78,000
Occupational Safety and Health Administration	$88,000
Environmental Protection Agency	$7,600,000

SOURCE: Tengs et al., "Five Hundred Life-Saving Interventions and Their Cost-Effectiveness," *Risk Analysis* 15, no. 3 (1995), pp. 369–90.

If the federal government shifted its resources from carrying out the highest-cost regulations to administering those that are more cost-effective, more lives would be extended for the same total cost. Alternatively, if the costliest regulations were loosened and the least costly were tightened, the same life-saving effects could be achieved at far lower cost.

Why are such programs and regulations implemented despite their inefficiency? One answer is that regulators are subject to the tunnel vision discussed in Chapter 4. Regulators recognize the potential benefits from stringent controls on the particular target for which they are responsible, but they tend to ignore the sacrifices that others must bear if the target is to be hit.

Regulators often see the goal as helping people by means of the regulators' specific programs rather than helping people by all government programs put together. Yet for the American people, who are presumably interested in the full complement of programs, excessive zeal for one program reduces the resources available to others and may be counterproductive.

Another reason for implementation of inappropriate regulations is selective and misleading communication about risks. George Gray and John Graham of the Harvard Center for Risk Analysis reviewed an EPA report on the risks caused by exposure to toxic air pollutants. They concluded that the EPA "misled journalists, policymakers, and the American people about what is known about the carcinogenic effects of certain air pollutants."[17] They found that prominent journalists, an important environmental leader, and even William Reilly, the EPA administrator at the time, had all misinterpreted the EPA's findings. Their interpretation reflected the report's summary rather than the more careful and accurate main narrative with its supporting data and calculations. The agency's misleading treatment of its own information fed the fears of citizens and encouraged demands for more stringent controls.

Such distortion leads voters to allow—and in some cases demand—regulatory programs that waste resources when other programs could have used those resources to save additional lives. The result has often been to expend large amounts of resources to achieve small marginal gains in risk reduction, as in the case of Superfund.

One reason voters tend to ignore the costs of environmental programs is that many mistakenly believe that corporations, not people, are paying the costs of reducing air pollutants or cleaning up chemical wastes. In fact, the cost of such regulations is spread among the firms' customers, employees, and shareholders. Fortunately, as more and more Americans invest in stock and thus own a significant number of corporate shares, voters may begin to realize that they are bearing a portion of the costs placed on business.

8. Market-like incentive schemes ("market-based mechanisms") may have benefits, but they are not the same as markets.

Economists often propose that the government address environmental problems by mimicking the private sector. It is now popular for economics textbooks to discuss pricing and market-like mechanisms for environmental policy. A few environmental groups have supported limited examples of such mechanisms.

These mechanisms became acceptable as environmental goals grew more ambitious and the cost of meeting those goals grew. Because of the high cost, elected officials and other policymakers began to recognize how charging companies for pollution and even trading rights in an artificial market for pollution permits could substantially reduce the cost of reaching environmental goals. There are two major approaches to "market mechanisms": pollution charges and trading systems.

Pollution charges are levied by the government on a polluter. Ideally, these charges would be equal to the costs borne by others downwind or downstream—thus giving the polluter an incentive to reduce the emissions in order to reduce the tax. As emissions went down, the payment by the company would go down—to the point at which the cost of more reduction would be larger than the further reduction in emission charges. At that point, the company would pay what remained of the tax. This process would result in more efficient emission controls.

In principle, revenue from the charges could be given as compensation to those downwind or downstream who actually bore the costs of the remaining pollution. In practice, however, this does not happen with existing pollution charge schemes in the United States.

In fact, only a few examples of pollution charges of this kind exist in the United States. The lack of such charges is probably due to lobbying pressure from companies that would pay because the charges would be an added cost of doing business. But there is an extra dimension: governments would not be happy with the charges either. The more effective companies became in controlling pollution, the less tax revenue there would be.

Emission charges have another problem: they do not allow for the flexibility of trades (see the next approach). Authorities announce the rules of behavior rather than establish tradable rights. Bargaining and trades among emitters and receptors seeking mutual benefits are not allowed.

Trading systems operate differently. The pollution control authority (such as the EPA) sets the total amount of emissions allowed in an area. Then it assigns (or perhaps sells) permits allowing emissions of that amount of pollution. These permits are tradable.

Tradable permits give emitters an incentive to reduce or even eliminate emissions. If they can reduce emissions cheaply, they may achieve enough reductions that they can sell their emissions permits to other emitters that face higher costs. In other words, some firms may not need all their permits, which may be valuable to others. Those firms that cannot reduce their emissions cheaply may consider buying permits.

Each polluting company will reduce pollution up to the point at which the added cost of reducing pollution further does not pay for itself. Some polluters (the ones that can reduce pollution cheaply) will reduce pollution as long as they can offset their costs by selling permits. Other polluters (for which reduction is more expensive) will reduce emissions to a point but then pay a company that can reduce emissions at a lower cost by buying some of their permits.

Efficiency in the attainment of the emission levels chosen by the pollution control authority should result. The best-known example is the

85

system of tradable permits to emit sulfur dioxide that was established in the Clean Air Act Amendments of 1990.

This act authorized the EPA to establish a nationwide program for trading sulfur dioxide emission reductions among power plants. The program is lowering the long-run compliance costs of electric utilities trying to meet legislated targets for those reductions. The utilities that face the highest cost of emission control are purchasing permits from those that can control more of their emissions at lower cost and thus have permits to sell.

The sulfur dioxide program was not the nation's first pollution credit trading program. During the 1980s, oil refiners were required to phase out lead in gasoline. Lead had been an important performance-enhancing additive in gasoline, but when emitted by automobiles, it dangerously polluted the air. Taking lead out of the gasoline meant that refiners had to reformulate their gasoline to get comparable performance without lead. Some could do it more easily than others.

Rather than simply demanding that each producer of gasoline take the lead out immediately, the EPA distributed permits on the basis of past production. The EPA allowed low-cost producers of leaded gasoline to purchase production permits from high-cost producers. High-cost producers stopped producing leaded gasoline sooner than a simpler set of command-and-control rules would have dictated. Low-cost producers took up the slack so that the cost of phasing out lead was minimized.

Such schemes can play a useful role, but they come into operation only after an inflexible level of total allowable emissions has been set. Each polluter is assigned rights to emit a certain amount. Then polluters can trade rights with one another. This trading should minimize the cost of reaching the agency-chosen standard, but it does not change the chosen standard.

Pollution control authorities need accurate information to set the allowable emissions at efficient levels in the first place. What level of emissions would produce a reduction in harm large enough to meet community standards of health and safety? Would a lower level avoid significant property damage to neighbors or to those downwind or downstream? Would a somewhat higher level of pollution, if allowed, still meet community standards?

Without this knowledge, the pollution control authority may worsen the situation rather than improve it. Yet these are exactly the same data that the courts need to enforce the rights of those harmed by pollution under the common-law system that has been largely rejected. When that information is not available, there is no reason to expect government regulation to do better than the courts in enforcing common law. And under current U.S. law, regulators usually have no obligation to ·seek such information.

Regulators have an additional problem that the courts do not. After a court orders a reduction in pollution to meet community standards against harm to others, the polluter and the victim can then trade property rights if they wish. For example, a court may order a factory owner to reduce air pollution by half to keep from violating the rights of a downwind farmer. However, the farmer might be willing to accept the pollution without the ordered reduction in exchange for $100,000. If avoiding the extra pollution control would save the polluter $200,000, then at some price in between both would gain. Efficient results would come about, even if the court's order—made without considering the cost of control—was not itself efficient.

With regulation, however, trade cannot usually follow, even though it would benefit all parties. Suppose the EPA set the same pollution standard as the judge did in the hypothetical case. The farmer downwind might prefer getting $100,000 rather than less pollution, and the factory owner might gladly pay that amount and more because reducing the pollution would cost more. But the two cannot legally make that trade under EPA regulations.

Even where trading is allowed, Congress or the EPA may set total allowable emissions so low that no amount of trading could reduce costs enough to make the benefits offset the costs. And regulators may not allow the pollution level to be relaxed.

Robert Crandall, a Brookings Institution economist, was a self-described "unabashed advocate of 'market solutions' to environmental problems" when he was a member of the Carter administration in the late 1970s. He championed programs such as emissions charges and tradable credits. Later, however, he became more skeptical.

"The emissions trading provision," he wrote, "was buried in a section of the Act that requires an annual 10 million ton rollback of SO_2 emissions." This enormous cutback was set even though the problem it was designed to combat—acid rain—had been shown to be "hardly the problem" it had previously been thought to be. "The costs are likely to swamp the benefits," Crandall said about the Clean Air Act Amendments of 1990.[18]

Thus, an efficient way to reach a badly chosen goal may be worse than no action at all. Any hope that incentive-based approaches will improve policy depends entirely on a proper choice of goals. The problems associated with politically determining the goals may completely overwhelm the benefits that come from market-like mechanisms.

9. Scientists, called on to evaluate the danger from a particular environmental concern, can be expected to focus attention on the most troublesome future scenarios that they can reasonably project.

Valid scientific information about environmental harms is not always easy to obtain. This is the case in problems ranging from chemical risks to global warming. Scientists often disagree on the severity of a particular problem, even when they agree on the basic science. An example is the debate over how serious a threat is posed by global warming that is caused by carbon dioxide emitted from the burning of fossil fuels. Issues that spark disagreement include the role of added carbon dioxide on the formation and composition of clouds, the health effects of warmer weather, and the impact that additional warming might have on sea level and polar ice caps.

One thing we can be sure of is that the scientists themselves—especially those in charge of large research projects and laboratories—have an incentive to seek more funding for their programs. Like all of us, they have tunnel vision regarding the importance of their missions. Each believes that his or her mission is exceedingly important relative to other budget priorities. (We will discuss this further in the final chapter on energy and global warming.)

When researchers seek to obtain more funding for their projects, having the public (and thus Congress and, potentially, private funders) worried about the critical nature of the problem being studied helps immensely. This incentive makes key researchers unlikely to interpret existing knowledge in a way that reduces public concern.

Heightening that concern helps the researcher. Whatever the evidence indicates, such scientists can be expected to emphasize the worst case that can reasonably be projected. "Scientists have realized that frightening the public brings state dollars," commented the late Sylvan Wittwer, a biologist from Michigan State University who had seen environmental crises come and go.[19]

10. Market solutions allow diverse decisions.

No one environmental plan can be best for everyone because, although many goals are shared, the emphasis on each differs among individuals. This idea parallels an important fact in nature: any change in the environment will help some plants and animals and will harm others. So far as is known, nature knows no favorites between one environment and another. Nature does not prefer forests to deserts or wetlands to prairies. Rather, each environment favors some living things over others.

As we seek prosperity, peace in society, and insurance against unforeseen environmental calamities, we should be aware of some major benefits of market solutions to environmental problems.

Market decisions are diverse and decentralized. Many mistakes will be made, but they will have far smaller effects than if central planners make them for the entire society. Those who disagree with a policy, such as the way the Nature Conservancy manages its lands, do not have to support those projects. Some ideas that seem mistaken at the time, such as Rosalie Edge's commitment to preserving predatory birds, can be greatly appreciated later.

Furthermore, the prosperity that the market system brings about fosters the willingness and the ability to seek and support ever-greater environmental quality.

Conclusion

Although private property rights promote prosperity and protect the environment, those who want more environmental protection frequently go to the government to obtain it.

Government regulation has had its successes. But government easily goes astray. Thus, we should consider protections against the misuse of government. They include the government budget process and legal rules that agencies must follow. The budget process will help prevent an agency's tunnel vision from taking too many resources from society's many other worthy goals. Legal rules, including requiring the agency to carry the burden of proof when exercising police powers, will keep it from harmful overreach.

The role of property rights and markets may face their biggest environmental challenge with today's biggest worry about the environment: climate change and the role of energy in today's economy.

That big topic is the subject of our final chapter.

6. Climate Change and the Problem of Energy

The Industrial Revolution, which began in England and Scotland in the late 18th century, is the name historians give to the second major economic transformation in world history (the first was the invention of agriculture 12,000 years earlier).

The forces unleashed by the Industrial Revolution enabled people to produce more, live longer, obtain greater leisure—and multiply themselves many times over. "After 1750 the fetters on sustainable economic growth were shaken off," says economist Joel Mokyr, and they were never to return.[1]

What made the Industrial Revolution revolutionary was its ability to harness energy. And energy remains one of the most important forces in economies today.

For thousands of years, people depended on human and animal power to cultivate the land, to produce clothing and shelter, and to move from place to place. There was improvement in people's lives, but it was slow.

The transformation began modestly, with the capture of wind and water energy, used initially to mill grain. But it gained speed when people began using coal to generate steam, thereby accelerating processes such as the production of cotton cloth and creating new modes of transportation such as steamships.

But new energy sources came with a price—both human and environmental. For example, coal had to be extracted from deep underground, where fatal accidents were common until the mid-20th century. Later, when coal was extracted from mines on the surface of mountains, the human toll was lighter, but the impact on the landscape was severe.

And when coal was used in production or for heat, it emitted smoke particulates and chemical byproducts. The famous "fog" of 19th-century London was caused by smoke from homes heated by coal.

The cleaner fuels that came along—oil and, later, natural gas—caused pollution too, although less. The smoke from gasoline combustion in automobiles in low-lying geographic basins such as Los Angeles and Denver created what became known as "smog."

In the United States, as we saw earlier, progress to combat pollution developed over time. Auto engines became much more efficient and less polluting; so did coal-fired power plants. Public pressure, laws and regulations, and companies' desire to stop losing smoke byproducts such as sulfur all contributed to improvement.

Two major energy-related issues are in the forefront of people's minds today. One is whether carbon-based fossil fuels are leading to severe increases in temperatures and other serious climatic effects. The other is whether we are running out of energy.

Let's look at them in turn.

1. Do fossil fuels cause global warming?

On a hot June day in 1988 at a congressional hearing in Washington, D.C., James E. Hansen, a scientist with the National Aeronautics and Space Administration (NASA), brought up the greenhouse effect. He told a committee that he had "99 percent confidence" that the world was getting warmer because of carbon dioxide and other "greenhouse" gases in the atmosphere.

We have been living with that concern ever since.

The theory underlying global warming goes back to a 19th-century chemist, Svante Arrhenius. He proposed that carbon dioxide and some other gases could create a blanket around the Earth that traps heat in the atmosphere—heat that would otherwise be radiated into space, keeping the Earth cooler. The theory was mostly an academic one until the late 20th century.

As the use of carbon-based fuels such as coal, oil, and natural gas accelerated, the amount of carbon dioxide in the atmosphere increased. There is fear that recent warm temperatures are caused by such gases, especially carbon dioxide, and that the Earth will continue to warm to dangerous levels.

The science is, of course, complex. Although no one disputes a very small greenhouse effect, extreme increases in temperature could occur only if carbon dioxide leads to increased clouds and water vapor, creating a major heat blanket. Figuring out whether that will happen, and to what extent, is the subject of debate.

The rhetoric has become heated, a situation which often happens when an issue becomes political. John Kerry, secretary of state under President Obama, has called climate change "the biggest challenge of all that we face right now."[2] Matt Ridley, a respected science writer who doubts that it is the biggest challenge facing the world, says that once he began to question the predictions of dire temperature changes,

> One by one, many of the most prominent people in the climate debate began to throw vitriolic playground abuse at me. I was "paranoid," "specious," "risible," "self-defaming," "daft," "lying," "irrational," an "idiot."[3]

A group of scientists organized by the United Nations and known as the Intergovernmental Panel on Climate Change (IPCC) has been issuing periodic reports that predict climate in the future. The latest report, issued in 2014, offers four different possible temperature projections by the end of the 21st century. Average temperatures could increase by as little as 0.3 degrees Celsius or as much as 4.8 degrees Celsius, compared to the period between 1986 and 2005. When compared with pre-industrial times, the IPCC says that average global temperatures could increase by either less than 2 degrees Celsius or more than 2 degrees Celsius—the difference dependent primarily to the extent to which carbon emissions are controlled.[4]

This reliance on a range of numbers and a focus on 2 degrees Celsius represents a more modest and less precise estimate than in the past. In 2001, the chairman of the IPCC flatly stated that the increase could be between 1.4 and 5.8 degrees Celsius.[5]

Furthermore, since 1998, there has been no warming trend. Certainly, global temperatures are high when compared with the past (records have been kept since the mid-1800s), but they are not increasing much from year to year, if at all.

This trend doesn't mean that warming has stopped permanently, but it does mean that the predictions have been off base. As *The Economist* stated, "the mismatch between rising greenhouse-gas emissions and not-rising temperatures is among the biggest puzzles in climate science right now."[6]

Nevertheless, the IPCC predicts that we can expect glaciers to melt, sea levels to rise, and weather extremes to increase. The IPCC notes that some of these things have already happened. From the latest summary for policymakers:

> It is *very likely* that heat waves will occur more often and last longer, and that extreme precipitation events will become more intense and frequent in many regions. The ocean will continue to warm and acidify, and global mean sea level to rise. [Emphasis in original.][7]

Not everyone agrees with these claims. For one thing, the temperature increase could be quite small. Furthermore, the computer models on which the IPCC projections are based have not been good at predicting recent and current temperatures. According to retired Massachusetts Institute of Technology scientist Richard Lindzen, the only way that past predictions have even roughly conformed to reality is through tempering the model by assuming some atmospheric cooling factors—such as the impact of aerosols from air pollution.[8]

Yet concern about greenhouse gases (which include methane and nitrous oxide as well as carbon dioxide) has led to enormous changes in government policy and to international negotiations to reduce emissions of carbon.

Does economics have something to tell us about climate change and about how to choose between government and market activity?

Yes.

Keep in mind that economics cannot answer the scientific questions surrounding climate change. Are temperatures going up over the long term? How do we know? If they are increasing, is it because of human action? What other natural forces are affecting temperatures? Those questions must be answered through scientific inquiry.

However, there is much that economics can offer, especially as we address the question of what the government should do and what private markets should do. Economics can tell us how people make decisions in the face of uncertainty, guided by incentives that reflect, in part, the narrow goals that all people have, as discussed earlier in this book.

So, consider incentives. It is to be expected that companies (such as electricity producers) that might be hit with regulations to reduce their carbon emissions would have an incentive to oppose such policies. As special interests, they might well use their political clout to prevent regulation (see Number 6 in Chapter 4).

Not so evident is the fact that scientists have incentives, too. They receive billions of dollars from the federal government to study climate change. According to the Government Accountability Office, in 2010 over $2 billion was spent on climate research and another $5 billion on projects to develop technology to cope with global warming.[9] Entire laboratories—government and private—depend heavily on federal funding for climate change–related studies.

So scientists may conduct their research objectively, but they have a strong incentive to focus on probable problems and to downplay good news; otherwise, their laboratories might suffer.

Others, too, face incentives that affect the global warming debate. Consider the media. We all know that bad news makes headlines ("if it bleeds it leads" is a common journalistic expression), whereas good news doesn't. As science writer Michael Fumento once wrote, no journalist is likely to write a headline that says, "Earth Not Destroyed; Billions Don't Die."[10] For this reason and others (the media tend to favor more government controls), there appears to be a bias in the media against favorable news about global warming.

Politicians have an incentive to adopt laws assuring the public that they are addressing global warming. As we saw earlier, laws are especially popular with politicians when their costs appear to fall on someone other than the voters (such as "industry") or if the costs are in the future.

Furthermore, politicians may receive campaign funds from special interests on both sides of the issue—from environmental groups that promote the regulations and companies that are trying to block the regulations or at least make them as palatable as possible. Such campaign funds provide politicians with an incentive to pay attention to global warming, even if nothing ultimately is done.

Ironically, it is widely recognized that any foreseeable policies, such as regulation of carbon dioxide emissions or a tax on carbon, will have only a tiny effect on global temperatures. When the United States was considering a treaty (the Kyoto Protocol) designed to bring down carbon emissions by 7 percent from the 1990 levels, scientist James Hansen said (in a paper with others) that the decrease would have "little effect in the 21st century."[11]

Actually reducing emissions of carbon dioxide into the atmosphere is a gargantuan task. Although cutting the use of fossil fuel would have some effect and more efficient production of fossil fuel might also, the Environmental Protection Agency (EPA) is promoting what it calls carbon dioxide capture and sequestration, or CCS, for coal-fired electric plants.

CCS means capturing the carbon dioxide as it is produced by combustion and piping it underground more than a mile deep. This complex process could push up electricity prices and reduce economic growth. (One estimate from the Congressional Research Service is that carbon capture would increase the cost of producing electricity at a new coal-fired power plant by 60 to 80 percent.)[12]

And yet there *has* been a decrease in carbon emissions in the United States—even before the EPA's most recent rules have taken effect. In 2015, the U.S. Department of Energy reported that emissions of carbon dioxide in the United States in 2014 were 9.7 percent lower than in 2007. Some of this decline was due to the recession that occurred in the United States starting in 2008, but the increasing use of natural gas, which when burned emits less carbon dioxide than coal or oil, was another big factor.[13]

Other nations are not likely to experience the same kind of reductions—especially developing nations. They want the lifestyle typical of Americans and Europeans and see economic growth through energy as the way to obtain that lifestyle.

But the tradeoffs are not just between economic growth now and possible reduction of temperatures decades in the future. There are also environmental tradeoffs.

Researcher Indur Goklany observes that some of the threats posed by global warming—hunger, malaria, water shortages, and coastal flooding—are *already* problems facing mankind. Why not address them now?[14]

And it is not at all clear that a warmer world would be a more dangerous world. As Goklany points out, the warming of the Earth since 1920, whatever the reason, has coincided with a global reduction in deaths from extreme events such as hurricanes, heat, drought, and so on by 98 percent.

Furthermore, a wealth of evidence suggests that it is better to have a strong economy that can adapt to problems as they arise, such as extreme weather events. A hurricane in Florida may cause severe damage, especially to buildings close to the shore, but it kills far fewer people than a hurricane in a poorer country such as Guatemala.

One of the messages of this book is that we cannot have a pristine environment without giving up other things. Furthermore, handing the job over to the government raises additional problems, such as the special-interest effect.

All those factors suggest that caution in promoting government policies that address global warming is in order. But climate change is not the only worry for environmentalists and others in the 21st century. The other is worry that we are running out of energy.

2. Are we running out of energy?

As far back as the 18th century, Thomas Malthus was concerned that the world would run out of food if people continued to produce children at a rapid rate.

In fact, that didn't happen. Birth rates went down dramatically once better nutrition and life-saving medicines were developed, because more children lived to maturity. And new technology led to almost unbelievable increases in food production.

The term "Malthusian" remains, however. Today, it is applied to concerns about running out not just of food but of all natural resources. Malthusian worries have always been in the background of the environmental movement—fear of overpopulation and inadequate food production as well as fear that the world is running out of sufficient land, water, and natural resources, including energy.

In 1972, a group of business executives and scientists called the Club of Rome issued a small book that was as alarming as anything James Hansen said 16 years later about global warming. On the basis of computer models (then still in their infancy), *Limits to Growth* claimed that the Earth was likely to deplete its natural resources, from minerals such as copper to energy resources such as oil and natural gas. The authors wrote:

> If the present growth trends in world population, industrialization, pollution, food production, and resource depletion continue unchanged, the limits to growth on this planet will be reached sometime within the next one hundred years.[15]

When the price of oil quadrupled in 1974, their worries seemed confirmed. Over the next few decades, however, those fears were not borne out by reality. Prices of energy fell in the 1980s, and concern about depleting natural resources subsided.

Economics explains why. When a resource becomes increasingly scarce, it becomes more expensive. That leads consumers to seek cheaper alternatives and spurs suppliers to seek new supplies, find substitutes, or recycle. The result tends to be a leveling of prices and less pressure on the resource.

This process has occurred many times, with many natural resources. Fears of crisis, usually reflected in rising prices, lead to actions that prevent a resource from disappearing. In the mid-19th century, for example, there was a crisis of whale oil, which was used for most artificial lighting. Whales were becoming harder to find on the high seas, and the price rose from $0.23 per gallon of oil in 1820 to $1.42 in 1850.

That led to a search for substitutes. At first, kerosene made out of coal oil took the place of whale oil, but then the gooey substance called "rock oil" was discovered in Pennsylvania and became the source of kerosene, as well as of the petroleum that would fuel so much economic growth in future years. (By the way, whale oil was still available, but by 1896 its price had sunk to 40 cents a gallon and hardly anyone wanted it.)[16]

Thus, economics and history combine to assure us that natural resources are unlikely to be depleted.

Over the years, the price and availability of energy have fluctuated enormously. In large part, fluctuations occurred because of geopolitical factors such as the oil embargo by the Organization of the Petroleum Exporting Countries (OPEC) in 1973 and the 1979 oil shock that stemmed from the fall of the Shah of Iran. More recently, the tragedy of September 11, 2001, had a fundamental effect on energy. The previously unimaginable attacks on the World Trade Center and the Pentagon that killed nearly 3,000 people changed Americans' view of energy, as well as of many other aspects of life.

The September 11 conspirators had come almost entirely from Saudi Arabia, which was at the time the world's largest oil producer. Their Middle Eastern origins aroused worries that the United States was too dependent on foreign oil—especially Middle Eastern oil. Indeed, the nation subsequently became embroiled in wars in Iraq and Afghanistan that many say were designed to keep the oil flowing.

Later, concern arose that Europe was too dependent on Russian oil, especially after relations with the West deteriorated under Vladimir Putin. Russia has been a leading oil exporter for many years.

Concerns about energy supply changed from natural resource depletion to whether geopolitical forces would prevent the United States from meeting its demand for energy.

Those concerns have been addressed in a most surprising way—although it is just the way that markets are supposed to work. In the first decade of the early 21st century, oil and natural gas production in the United States and Canada began to surge.

In 2014, the United States became the world's largest producer of oil (including liquids from natural gas), outperforming Russia and Saudi Arabia. The achievement was due to market forces and technology rather than to any specific government policy.

As Robert Bryce, a science writer with the Manhattan Institute described it, the change was "the result of a century of improvements to older technologies such as drill rigs and drill bits, along with better seismic tools, advances in materials science, better robots, more capable submarines, and, of course, cheaper computer power."[17]

The displacement of Saudi Arabia and Russia as the top oil producers took place only four years after *Science* magazine quoted a government oil analyst:

> All in all, "technology matters, economics matters, but geology really does matter," says oil analyst David Greene of the U.S. Department of Energy's Oak Ridge National Laboratory in Tennessee. "Progress in technology is not fast enough to keep up with depletion" of oil reservoirs.[18]

But it turns out that technology and economics did keep up.

Among the technologies that have improved dramatically are directional drilling and hydraulic fracturing (or "fracking"). Directional (or horizontal) drilling allows producers to find oil and gas over a broader area. Fracking involves shooting water, sand, and chemicals deep into shale rock or "tight sands" at such high pressure that the shale breaks, thus opening up areas that have natural gas or oil.

It took little time for fracking to come under fire from environmentalist critics, however. The main focus has been "flowback" water. After the sand, water, and chemical mixture has broken the rock deep underground, some of that mixture returns to the surface or near the surface. In some cases, the mixture may have contaminated groundwater, and in early cases, some of it was dumped into streams. Tim Fitzgerald, an economics professor at Montana State University, explains that some of the chemicals in the mixture are toxic, even though they are extremely dilute. Companies are reluctant to reveal what they are because they consider the information a trade secret.[19]

In Pavillion, Wyoming, claims of groundwater contamination brought in the EPA, which conducted tests between March 2009 and April 2011. The agency did find contamination, but a subsequent study by the U.S. Geological Survey discredited the EPA results, and the agency never produced a final report.

However, as Fitzgerald observes, the issue continues to simmer, in Pavillion as elsewhere—and rightly so. As a new technology—or, more accurately, a new suite of old and evolving technologies—is tried out, mistakes will be made. The property rights of groundwater owners as well as other owners, such those who own other underground shale, should be respected.

Fracking raises one of the fundamental issues of this book: to what extent can these problems be worked out by property owners working together or using common-law courts? To what extent should the government interfere?

So far, it appears that both have happened. The initial problems, such as too-little distance between fracking and groundwater, have led private companies to change their practices. However, the government has also weighed in as a regulator. Indeed, at the time of this writing, the state of New York bans fracking, even though the neighboring state of Pennsylvania, which allows it, has experienced a boom in natural gas and oil production.

All in all, the increases in oil and gas production have been stunning. "The rapid growth in U.S. oil production has surprised even industry insiders," reported Patti Domm of CNBC in 2013. "Forecasts that once sounded far-fetched are becoming reality. The oil production boom had been expected, but the magnitude of change in such a short period of time is a surprise."[20] The change in energy production started with natural gas—reflecting the initial use of fracking—but then "unconventional" oil began to flow, primarily from shale and tight sands on private and state land.

To a large extent, the federal government sat out this boom. The Congressional Research Service reports that oil production on federal land and water declined by 6.2 percent between 2009 and 2013, and federal natural gas production fell by 27.8 percent during the same period.[21] And that was before the federal government issued new regulations on fracking in 2015. The report's author, Marc Humphries, suggests that the complex permitting process makes federal land less attractive than private and state land.

Conclusion

The discovery that private innovation and market activity have transformed the United States—often considered a "played out" source—into an energy powerhouse provides an appropriate conclusion to this chapter and to this book.

But before I go, let me repeat the theme. Both private markets and governments have a role in our country's activities, including those protecting the environment. The weight of opinion tends to push toward a greater role for government, even though that role is often misused and sometimes has unfortunate consequences. Economics shows us the wisdom of considering a greater role for market solutions.

So should government control and involvement in environmental regulations be shrunk? Should the role of the private sector be expanded? If the answer is yes, how can that happen?

Such questions have been addressed in this book with principles and examples but not with a cookbook of specific recipes. And that is the proper role of economics—helping us think through each problem more insightfully so that we can apply our own values and beliefs more effectively and better understand the arguments of others. I encourage all readers to do just that.

Notes

Chapter 1

1. Hank Fischer, "Who Pays for Wolves?," *PERC Reports* 19, no. 4 (2001), http://www.perc.org/articles/who-pays-wolves.

2. Rosalie Edge's life (including these anecdotes) is recounted in Dyana Z. Furmansky, *Rosalie Edge, Hawk of Mercy: The Activist Who Saved Nature from the Conservationists* (Athens: University of Georgia Press, 2010).

3. Robert H. Nelson, *The New Holy Wars: Economic Religion vs. Environmental Religion* (University Park: Pennsylvania State University Press, 2010), p. 13.

4. John Roach, "Earth Day Facts," *National Geographic News*, April 22, 2012, http://news.nationalgeographic.com/news/2012/04/120420-earth-day-facts-2012-environment-science-nation/.

5. Quoted in Jane S. Shaw, "Business and the Environment: Is There More to the Story?," *Business Economics* 40, no. 1 (2005): 40–45, http://www.perc.org/articles/business-and-environment.

6. World Commission on Environment and Development, *Report of the World Commission on Environment and Development: Our Common Future* (Oslo: United Nations, 1987), p. 16, http://www.un-documents.net/our-common-future.pdf.

Chapter 2

1. William Booth, "Developers Wish Rare Fly Would Buzz Off: Flower-Loving Insect Becomes Symbol for Opponents of Endangered Species Act," *Washington Post*, April 4, 1997.

2. Patrick J. Kiger, "Federal Study Highlights Spike in Eagle Deaths at Wind Farms," *The Great Energy Challenge* (blog), National Geographic, September 12, 2013, http://energyblog.nationalgeographic.com/2013/09/12/federal-study-highlights-spike-in-eagle-deaths-at-wind-farms/.

3. Todd Woody, "Solar Energy Faces Tests on Greenness," *New York Times*, February 24, 2011, p. B1, http://www.nytimes.com/2011/02/24/business/energy-environment/24solar.html?_r=0.

4. William F. Hosford and John L. Duncan, "The Aluminum Beverage Can," *Scientific American*, September 1994, pp. 48–53.

5. More details about the Lucas case can be found in James R. Rinehart and Jeffrey J. Pompe, "The Lucas Case and the Conflict over Property Rights," in *Land Rights: The 1990s' Property Rights Rebellion*, ed. Bruce Yandle (Lanham, MD: Rowman & Littlefield, 1995), pp. 67–101.

6. Reed Watson, "Scott River Water Trust: Improving Stream Flows the Easy Way," PERC Case Study, Property and Environment Research Center, Bozeman, MT, December 2013, http://www.perc.org/sites/default/files/pdfs/Final-Scott%20River %20Water%20Trust.pdf.

7. Lynn Scarlett, "New Environmentalism," NCPA Policy Report 201, National Center for Policy Analysis, Dallas, TX, January 1997, p. 11.

8. Pierre Desrochers, "The Secret Past of Resource Recovery," *PERC Reports* 17, no. 3 (1999), http://www.perc.org/articles/secret-past-recycling.

9. Gene M. Grossman and Alan B. Krueger, "Economic Growth and the Environment," *Quarterly Journal of Economics* 110, no. 2 (1995): 353–77.

10. Don Coursey, "The Demand for Environmental Quality," working paper, John M. Olin School of Business, Washington University, Saint Louis, MO, December 1992.

11. Sierra Media Group, "Reader Profile," June 2014, http://www.sierraclub.org /sierra/media-kit/audience/print/.

12. U.S. Census Bureau, "USA QuickFacts," 2009–13 data, http://quickfacts.census.gov /qfd/states/00000.html.

Chapter 3

1. For more about Bourland's project, see Terry L. Anderson and Donald R. Leal, *Enviro-Capitalists: Doing Good While Doing Well* (Lanham, MD: Rowman & Littlefield, 1997), pp. 4–8.

2. As quoted in Will Durant, *The Life of Greece* (New York: Simon and Schuster, 1939), p. 536.

3. See Roger Meiners and Bruce Yandle, "The Common Law: How It Protects the Environment," PERC Policy Series PS-13, Property and Environment Research Center, Bozeman, MT, May 1998, pp. 4–10.

4. For the full story of the Rainey Preserve, see John Baden and Richard Stroup, "Saving the Wilderness: A Radical Proposal," *Reason* 13 (1981): 28–36; Pamela Snyder and Jane S. Shaw, "PC Oil Drilling in a Wildlife Refuge," *Wall Street Journal*, September 7, 1995; and John Flicker, "Don't Desecrate the Arctic Refuge," *Wall Street Journal* (letter), September 18, 1995.

5. Jen DeGregorio, "Audubon Sanctuary Considers Allowing Oil and Gas Drilling," *Times-Picayune*, April 26, 2012. http://www.nola.com/business/index.ssf/2010/01 /audubon_society_sanctuary_cons.html.

6. David Yarnold, "Audubon View," *Audubon*, July–August 2012, http:// www.audubon.org/magazine/july-august-2012/audubon-view.

7. Kurt Repanshek, "Is Outsourcing Parks a Key to Solving the National Park Service's Financial Woes?," *National Parks Traveler*, October 12, 2014, http:// www.nationalparkstraveler.com/2014/10/outsourcing-parks-key-solving-national -park-services-financial-woes25723?page=1.

8. Mark Duda and Jane S. Shaw, "A New Environmental Tool? Assessing Life Cycle Assessment," Contemporary Issues 81, Center for the Study of American Business, Saint Louis, MO, 1996.

9. Rich Ceppos, "The Car Who Came in from the Cold," *Car and Driver*, December 1990, pp. 89–97.

10. Scott Shane, "Start Up Failure Rates: The Definitive Numbers," *Small Business Trends,* December 17, 2012, http://smallbiztrends.com/2012/12/start-up-failure-rates-the-definitive-numbers.html.

11. Anderson and Leal, *Enviro-Capitalists,* pp. 75–77.

Chapter 4

1. Hernando de Soto, *The Mystery of Capital* (New York: Basic Books, 2000).

2. Stephen Breyer, *Breaking the Vicious Circle: Toward Effective Risk Regulation* (Cambridge, MA: Harvard University Press, 1993), pp. 11–19.

3. Bruce Yandle, "Markets for Water Quality," *PERC Reports* 26, no. 3 (2008), http://www.perc.org/articles/markets-water-quality#sthash.9V4zuC9l.dpuf.

4. Many people still consider the Love Canal incident an example of pollution by a business because the waste site had been built by a chemical company. In fact, however, the problems came about because the local school board forced the company to sell the site (for $1) to be used for a school. In spite of extensive warnings by the company, the board later let some of the land be sold for development. See Eric Zuesse, "Love Canal: The Truth Seeps Out," *Reason* 12 (1981): 17–31.

5. James T. Hamilton and W. Kip Viscusi, *Calculating Risks* (Cambridge, MA: MIT Press, 1999).

6. Frank Partnoy, "The Cost of a Human Life, Statistically Speaking," *Globalist*, July 21, 2012, http://www.theglobalist.com/the-cost-of-a-human-life-statistically-speaking/.

7. Rudy Abramson, "The Superfund Cleanup: Mired in Its Own Mess," *Los Angeles Times,* May 10, 1993, http://articles.latimes.com/1993-05-10/news/mn-33610_1_superfund-reform.

8. Stephen Moore and Joel Griffith, "Media, Environmentalists Were Wrong: How the Gulf Coast Roared Back after Oil Spill," *Daily Signal*, April 25, 2015, http://dailysignal.com/2015/04/25/media-environmentalists-were-wrong-how-the-gulf-coast-roared-back-after-oil-spill/.

9. These figures come from Terry L. Anderson and Pamela S. Snyder, *Water Markets: Priming the Invisible Pump* (Washington, D.C.: Cato Institute, 1997), p. 10.

10. More information about special interests can be found in Chapter 6 of James D. Gwartney et al., *Economics: Private and Public Choice*, 15th ed. (Independence, KY: South-Western/CengageBrain, 2015).

11. Juliet Eilperin, "Environmentalists Take Hard Line with Obama on Keystone XL," *Washington Post*, September 24, 2013, http://www.washingtonpost.com/blogs/post-politics/wp/2013/09/24/environmentalists-warn-obama-against-keystone-xl-even-if-canada-compromises-on-climate/.

12. Louis Jacobson, "Only 1 Percent of Endangered Species List Have Been Taken Off List, Says Cynthia Lummis," Politifact.com (*Tampa Bay Times*), September 3, 2013, http://www.politifact.com/truth-o-meter/statements/2013/sep/03/cynthia-lummis/endangered-species-act-percent-taken-off-list/.

13. From the transcript of a talk by Michael Bean at a U.S. Fish and Wildlife seminar, November 3, 1994, Marymount University, Arlington, VA.

14. Lee Ann Welch, "Property Rights Conflicts under the Endangered Species Act: Protection of the Red-Cockaded Woodpecker," PERC Working Paper 94-12, Property and Environment Research Center, Bozeman, MT, 1994.

15. Ike C. Sugg, "Ecosystem Babbitt-Babble," *Wall Street Journal* (commentary), April 2, 1993.

16. Welch, "Property Rights Conflicts," p. 47.

17. Susan McGrath, "Let's Make a Deal," *Audubon*, January–February 2008, http://www.audubon.org/magazine/january-february-2008/lets-make-deal.

18. Jay Newton-Small, "Don't Cry for K Street: Federal Lobbying Is Down, but Profits Are Up," *Time*, August 8, 2013, http://swampland.time.com/2013/08/08/dont-cry-for-k-street-federal-lobbying-is-down-but-profits-are-up/.

19. Clyde Wayne Crews, *Ten Thousand Commandments 2014* (Washington, D.C.: Competitive Enterprise Institute, 2014), https://cei.org/studies/ten-thousand-commandments-2014.

20. Angela Logomasini, "The Green Regulatory State," Issue Analysis 9, Competitive Enterprise Institute, Washington, D.C., August 2007, http://www.cei.org/pdf/6106.pdf.

21. Bruce Yandle, "Bootleggers and Baptists in Retrospect," *Regulation* 22, no. 3 (1999): 5–7.

22. Kate Galbraith, "Threatened Smelt Touches Off Battles in California's Endless Water Wars," *New York Times*, February 14, 2015. http://www.nytimes.com/2015/02/15/us/threatened-smelt-touches-off-battles-in-californias-endless-water-wars.html?_r=0. See also "California's Man-Made Drought: The Green War against San Joaquin Valley Farmers," *Wall Street Journal* (editorial), September 2, 2009, http://www.wsj.com/articles/SB10001424052970204731804574384731898375624.

23. Debra S. Knopman, "Pennywise, Billions Foolish: The Folly of Underinvestment in Environmental Monitoring," originally published by IntellectualCapital.com and cited in Richard A. Halpern, "The New Watershed Tools: Genuine Steel or Chrome-Plated Plastic?" *Water Resources Impact* 2, no. 6 (2000): 23–26.

Chapter 5

1. A hectare is 2.471 acres. The dollar value of crop production is expressed in 2004–6 international dollars. Figures are from Food and Agricultural Organization's FAOSTAT database (FAO). For Ethiopia, see http://faostat.fao.org/CountryProfiles/Country_Profile/Direct.aspx?lang=en&area=238. For the Netherlands, see http://faostat.fao.org/CountryProfiles/Country_Profile/Direct.aspx?lang=en&area=150.

2. James D. Gwartney, Robert Lawson, and Joshua Hall, *Economic Freedom of the World 2015 Annual Report* (Vancouver, BC: Fraser Institute, 2015), http://www.freetheworld.com/2015/economic-freedom-of-the-world-2015.pdf.

3. Gwartney, Lawson, and Hall, *Economic Freedom of the World*, p. 23.

4. Based on national income media figures, weighted by population.

5. Robert Rector and Rachel Sheffield, "The War on Poverty after 50 Years," Backgrounder 2955, Heritage Foundation, Washington, D.C., September 15, 2014.

6. Martin Malia, *The Soviet Tragedy: A History of Socialism in Russia, 1917–1991* (New York: The Free Press, 1994), p. 199.

7. David Osterfeld, *Prosperity versus Planning: How Government Stifles Economic Growth* (New York: Oxford University Press, 1992), p. 82.

8. These examples come from Peter J. Hill, "Environmental Problems under Socialism," *Cato Journal* 12, no. 2 (1992): 321–35.

9. Mark Hertsgaard, *Earth Odyssey: Around the World in Search of Our Environmental Future* (New York: Broadway Books, 1999), p. 3.

10. Mikhail Bernstam, *The Wealth of Nations and the Environment* (London: Institute of Economic Affairs, 1991), pp. 23–25.

11. Seth W. Norton, "Property Rights, the Environment, and Economic Well-Being," in *Who Owns the Environment?*, ed. Peter J. Hill and Roger E. Meiners, pp. 37–54 (Lanham, MD: Rowman & Littlefield, 1998).

12. Seth W. Norton, "Poverty, Property Rights, and Human Well-Being: A Cross-National Study," *Cato Journal* 18, no. 2 (1998): 233–45.

13. See Neil K. Komesar, *Imperfect Alternatives* (Chicago: University of Chicago Press, 1994). Chapters 1 and 2 of Komesar's book provide a full explanation of the nature of this institutional choice question.

14. Roger E. Meiners, Stacie Thomas, and Bruce Yandle, "Burning Rivers, Common Law, and Institutional Choice for Water Quality," in *The Common Law and the Environment: Rethinking the Statutory Basis for Modern Environmental Law*, ed. Roger E. Meiners and Andrew P. Morriss (Lanham, MD: Rowman & Littlefield, 2000), pp. 54–85.

15. See Elizabeth Brubaker, *Property Rights in the Defence of Nature* (Toronto, ON: Earthscan, 1995), p. 161.

16. Tammy O. Tengs, et al., "Five Hundred Life-Saving Interventions and Their Cost-Effectiveness," *Risk Analysis* 15, no. 3 (1995): 369–90.

17. George M. Gray and John D. Graham, "Risk Assessment and Clean Air Policy," *Journal of Policy Analysis and Management* 10, no. 2 (1991): 286–95, at p. 286. Gray and Graham analyzed a 1989 EPA publication, Office of Air Quality Planning and Standards, *Cancer Risk from Outdoor Exposure to Air Toxics* (Research Triangle Park, NC: Environmental Protection Agency).

18. Robert W. Crandall, "Is There Progress in Environmental Policy?" *Contemporary Economic Policy* 12, no. 1 (1994): 80–83.

19. Telephone conversation with Sylvan Wittwer, June 26, 2001.

Chapter 6

1. Joel Mokyr, "Editor's Introduction: The New Economic History and the Industrial Revolution," in *The British Industrial Revolution*, ed. Joel Mokyr (Boulder, CO: Westview Press, 1999), pp. 1–84, at p. 83.

2. Patrick Goodenough, "Kerry: Climate Change 'Biggest Challenge of All That We Face Right Now,'" *CNS News*, August 14, 2014.

3. Matt Ridley, "My Life as a Lukewarmer," *Matt Ridley's Blog*, January 20, 2015, http://www.rationaloptimist.com/blog/my-life-as-a-climate-lukewarmer.aspx.

4. Intergovernmental Panel on Climate Change, "Climate Change 2014 Synthesis Report: Summary for Policymakers," IPCC, Geneva, 2014, p. 10, http://www.ipcc.ch/pdf/assessment-report/ar5/syr/AR5_SYR_FINAL_SPM.pdf.

5. Bert Metz, "Statement on Behalf of the Chairman of the IPCC, Dr. Robert Watson," Seventh Conference of the Parties to the United Nations Framework Convention on Climate Change," Marrakesh, Morocco, November 7, 2001, https://www.ipcc.ch/graphics/speeches/robert-watson-november -2001.pdf.

6. *The Economist*, "A Sensitive Matter," March 30, 2013, http://www.economist.com/news/science-and-technology/21574461-climate-may-be-heating-up-less-response-greenhouse-gas-emissions.

7. IPCC, "Climate Change 2014 Synthesis Report: Summary for Policymakers," p. 10.

8. Richard Lindzen, "The Climate Science Isn't Settled: Confident Predictions of Catastrophe Are Unwarranted," *Wall Street Journal,* November 30, 2009, http://www.wsj.com/articles/SB10001424052748703939404574567423917025400.

9. Government Accountability Office, *Climate Change: Improvements Needed to Clarify National Priorities and Better Align Them with Federal Funding Decisions,* GAO-11-317 (Washington, D.C.: GAO, May 20, 2011), http://www.gao.gov/products/GAO-11-317.

10. Michael Fumento, *Science under Siege: Balancing Technology and the Environment* (New York: William Morrow, 1993), p. 342.

11. James E. Hansen, et al., "Global Warming in the 21st Century: An Alternative Scenario," NASA Goddard Institute for Space Studies, New York, April 17, 2001, p. 1.

12. Peter Folger, "Carbon Capture: A Technology Assessment," CRS Report R41325, Congressional Research Service, Washington, D.C., November 13, 2013, p. 18, https://www.fas.org/sgp/crs/misc/R41325.pdf.

13. U.S. Energy Information Administration, "Carbon Dioxide Emissions from Energy Consumption by Source," *Monthly Energy Review,* April 2015, http://www.eia.gov/totalenergy/data/monthly/pdf/sec12.pdf.

14. Indur M. Goklany, "Economic Development in Developing Countries: Advancing Human Well-Being and the Capacity to Adapt to Global Warming," in *Climate Coup: Global Warming's Invasion of Our Government and Our Lives,* ed. Patrick J. Michaels (Washington, D.C.: Cato Institute, 2011) pp. 157–85.

15. Donella H. Meadows, Dennis L. Meadows, and William W. Behrens, *The Limits to Growth: A Report for the Club of Rome's Project on the Predicament of Mankind* (New York: Universe Books, 1972), p. 23.

16. Gwartney et al., *Economics,* p. 681.

17. Robert Bryce, "New Technology for Old Fuels: Innovation in Oil and Natural Gas Production Assures Future Supplies," Energy Policy and the Environment Report 12, Manhattan Institute, New York, April 2013, p. i.

18. Richard A. Kerr, "Peak Oil Production May Already Be Here," *Science* 331, no. 6024 (2011): 1510–11, at p. 1510.

19. Tim Fitzgerald, "Frackonomics: The Economics behind America's Shale Revolution," *PERC Reports* 33, no. 2 (2014), http://www.perc.org/articles/frackonomics.

20. Patti Domm, "U.S. Oil and Gas Boom Takes Many by Surprise," CNBC, March 4, 2013, http://www.cnbc.com/id/100513916.

21. Marc Humphries, "U.S. Crude Oil and Natural Gas Production in Federal and Non-federal Areas," CRS Report R42432, Congressional Research Service, Washington, D.C., April 10, 2014.

Index

Page references followed by f indicate figures; n designates a numbered note.

About the Author

Richard L. Stroup is an emeritus economics professor at Montana State University and North Carolina State University. He spent most of his career at Montana State University and was affiliated with the Property and Environment Research Center (PERC) in Bozeman, Montana. His research helped develop the approach to resource problems often termed "new resource economics," which is also known as "free-market environmentalism."

Stroup is a long-time coauthor with James Gwartney of a leading economics principles textbook, *Economics: Private and Public Choice*, now in its 15th edition, with coauthors Russell S. Sobel and David Macpherson. He is a coauthor of *Common Sense Economics: What Everyone Should Know about Wealth and Prosperity*, with James Gwartney, Dwight Lee, Tawni Ferrarini, and Joseph Calhoun. He received his B.A., M.A., and Ph.D. degrees from the University of Washington. From 1982 to 1984, Dr. Stroup was director of the Office of Policy Analysis at the U.S. Department of the Interior. He lives in Raleigh, North Carolina, with his wife, Jane Shaw Stroup.

Cato Institute

Founded in 1977, the Cato Institute is a public policy research foundation dedicated to broadening the parameters of policy debate to allow consideration of more options that are consistent with the principles of limited government, individual liberty, and peace. To that end, the Institute strives to achieve greater involvement of the intelligent, concerned lay public in questions of policy and the proper role of government.

The Institute is named for Cato's Letters, libertarian pamphlets that were widely read in the American Colonies in the early 18th century and played a major role in laying the philosophical foundation for the American Revolution.

Despite the achievement of the nation's Founders, today virtually no aspect of life is free from government encroachment. A pervasive intolerance for individual rights is shown by government's arbitrary intrusions into private economic transactions and its disregard for civil liberties. And while freedom around the globe has notably increased in the past several decades, many countries have moved in the opposite direction, and most governments still do not respect or safeguard the wide range of civil and economic liberties.

To address those issues, the Cato Institute undertakes an extensive publications program on the complete spectrum of policy issues. Books, monographs, and shorter studies are commissioned to examine the federal budget, Social Security, regulation, military spending, international trade, and myriad other issues. Major policy conferences are held throughout the year, from which papers are published thrice yearly in the *Cato Journal*. The Institute also publishes the quarterly magazine *Regulation*.

In order to maintain its independence, the Cato Institute accepts no government funding. Contributions are received from foundations, corporations, and individuals, and other revenue is generated from the sale of publications. The Institute is a nonprofit, tax-exempt, educational foundation under Section 501(c)3 of the Internal Revenue Code.

CATO INSTITUTE
1000 Massachusetts Ave., N.W.
Washington, D.C. 20001
www.cato.org